Ecological Studies, Vol. 94

Analysis and Synthesis

Ecological Studies

Volume 73
Toxic Organic Chemicals in Porous Media (1989)
Z. Gerstl, Y. Chen, U. Mingelgrin, and B. Yaron (Eds.)

Volume 74
Inorganic Contaminants in the Vadose Zone (1989)
B. Bar-Yosef, N.J. Barnow, and J. Goldshmid (Eds.)

Volume 75
The Grazing Land Ecosystems of the African Sahel (1989)
H.N. Le Houérou

Volume 76
Vascular Plants as Epiphytes: Evolution and Ecophysiology (1989)
U. Lüttge (Ed.)

Volume 77
Air Pollution and Forest Decline: A Study of Spruce (*Picea abies*) on Acid Soils (1989)
E.-D. Schulze, O.L. Lange, and R. Oren (Eds.)

Volume 78
Agroecology: Researching the Ecological Basis for Sustainable Agriculture (1990)
S.R. Gliessman (Ed.)

Volume 79
Remote Sensing of Biosphere Functioning (1990)
R.J. Hobbs and H.A. Mooney (Eds.)

Volume 80
Plant Biology of the Basin and Range (1990)
B. Osmond, G.M. Hidy, and L. Pitelka (Eds.)

Volume 81
Nitrogen in Terrestrial Ecosystems: Questions of Productivity, Vegetational Changes, and Ecosystem Stability (1990)
C.O. Tamm

Volume 82
Quantitative Methods in Landscape Ecology: The Analysis and Interpretation of Landscape Heterogeneity (1990)
M.G. Turner and R.H. Gardner (Eds.)

Volume 83
The Rivers of Florida (1990)
R.J. Livingston (Ed.)

Volume 84
Fire in the Tropical Biota: Ecosystem Processes and Global Challenges (1990)
J.G. Goldammer (Ed.)

Volume 85
The Mosaic-Cycle Concept of Ecosystems (1991)
H. Remmert (Ed.)

Volume 86
Ecological Heterogeneity (1991)
J. Kolasa and S.T.A. Pickett (Eds.)

Volume 87
Horses and Grasses: The Nutritional Ecology of Equids and Their Impact on the Camargue (1992)
P. Duncan

Volume 88
Pinnipeds and El Niño: Responses to Enviromental Stress (1992)
F. Trillmich and K.A. Ono (Eds.)

Volume 89
Plantago: A Multidisciplinary Study (1992)
P.J.C. Kuiper and M. Bos (Eds.)

Volume 90
Biogeochemistry of a Subalpine Ecosystem: Loch Vale Watershed (1992)
J. Baron (Ed.)

Volume 91
Atmospheric Deposition and Forest Nutrient Cycling (1992)
D.W. Johnson and S.E. Lindberg (Eds.)

Volume 92
Landscape Boundaries (1992)
A.J. Hansen and F. di Castri (Eds.)

Volume 93
Fire in South African Mountain Fynbos (1992)
B.W. van Wilgen, D.M. Richardson, F.J. Kruger, and H.J. van Hensbergen (Eds.)

Felix Bärlocher (Ed.)

The Ecology of Aquatic Hyphomycetes

With 23 Figures in 107 Parts
and 18 Tables

Springer-Verlag
Berlin Heidelberg New York
London Paris Tokyo
Hong Kong Barcelona
Budapest

Professor Dr. *Felix Bärlocher*
Department of Biology
Mount Allison University
Sackville, N.B., EOA 3CO
Canada

ISBN 3-540-54400-3 Springer-Verlag Berlin Heidelberg New York
ISBN 0-387-54400-3 Springer-Verlag New York Berlin Heidelberg

Library of Congress Cataloging-in-Publication Data
The Ecology of aquatic hyphomycetes/Felix Bärlocher (ed.). p. cm. –
(Ecological studies; vol. 94)
Includes bibliographical references and index.
ISBN 3-540-54400-3 (Berlin: alk. paper): DM188.00 –
ISBN 0-387-54400-3 (New York: alk. paper)
1. Hyphomycetes – Ecology. 2. Aquatic fungi – Ecology.
1. Bärlocher, Felix. II. Series: Ecological studies; v. 94.
QK625.A1E36 1992 589.2'4 – dc20 92-9577 CIP

Printed in Germany

Production Editor: Martin Langner
Typesetting: Best-set Typesetter Ltd., Hong Kong
31/3145 5 4 3 2 1 0 Printed on acid-free paper

Foreword

Some of the favourite comments of reviewers of multi-authored books are that the various sections are uneven in quality, or incompletely integrated, or, that an overall theme is lacking. A seamless, complete story often seems to be the ideal. This may be appropriate for an undergraduate textbook; in my view, it does not convey a true picture of a field attractive to researchers. To paraphrase Simberloff's (1980) remarks on ecology: what reviewers view as noise, is music to the scientist; the individuality of approaches is the most striking, intrinsic, and inspiring characteristic of books reviewing research in progress. To reflect some of this diversity, I have invited respected, active scientists as co-authors. My main concerns have been to avoid obvious errors, and to keep overlap between individual chapters at a reasonable level. Nevertheless, each chapter is meant to be intelligible by itself, without constant reference to the rest of the book. The aim was to give as complete a review of the biology and ecology of aquatic hyphomycetes as possible. Obvious gaps remain; I hope these are due to the fact that no information is available, and not to an oversight of mine.

It is inevitable that researchers are influenced by the "contemporary intellectual milieu". Stream ecology, theoretical ecology and general fungal ecology (of soil fungi and, more recently, of wood-decomposing macrofungi) have a longer history and have been more influential than the studies of aquatic hyphomycetes. A critical evaluation of what these areas have to offer seemed appropriate. One approach would have been to ask external experts for a contribution. I decided instead to give my own interpretation, because I think that a thorough knowledge of the specific group should come first. Undoubtedly, others will reach different conclusions concerning the usefulness of some of the recent developments in stream ecology, theoretical ecology and fungal ecology. There is an abundance of books, reviews and symposia on these topics allowing readers to pursue these topics in more depth.

Ecologists often express the conviction that insights from their own field have general implications; conversely, they believe their field is unique and will therefore offer unique insights. Or, they claim that their field has great practical implications. Even though I have used similar sentences myself, I no longer find them particularly appropriate. They often have little to do with what attracts scientists to one field or another, and they are poor guides to where the next breakthrough will occur. Prospects for advancing our knowledge of aquatic hyphomycetes remain brightest if mycologists continue to feel the love of, and enthusiasm for, aquatic hyphomycetes still evident in C.T. Ingold. I don't doubt that readers of this book wishing to initiate research on aquatic hyphomycetes will be able to formulate their own rationalizations to convince granting agencies or prospective employers.

Many friends and colleagues have contributed, directly or indirectly, to this book. Above all, I'm grateful to Bryce Kendrick, who introduced me to aquatic hyphomycetes and who has continued to give support and advice whenever needed. I wish to thank Professor O.L. Lange for his enthusiastic support of this project. I'm grateful to my co-authors, who, for the most part, delivered their manuscripts on time. In addition, several of them kindly reviewed individual chapters, as did Maurice Lock and Mike Winterbourn. Ken Thomas, H.Q.P. Crick, C.T. Ingold, John Webster, Ludmila Marvanová, E. Descals and H.B.N. Hynes provided me with unpublished information. I'm particularly indebted to John Webster for his hospitality during my sabbatical year in Exeter, and for access to his laboratory and bibliography.

A.D.M. Rayner and C.A. Shearer, and the University of Cambridge Press and the New York Botanical Garden gave permission to reprint Fig. 3.1.

Aquatic hyphomycetes came into their own largely through the contributions of three scientists: C.T. Ingold, who discovered them; John Webster, who worked out much of their biology and ecology; H.B.N. Hynes, who emphasized the role of leaves in the energy flow of streams, and who established the role of fungi as link between leaves and invertebrates. I therefore dedicate this book to C.T. Ingold, John Webster and H.B.N. Hynes.

Sackville/Canada, 30 May, 1991 *Felix Bärlocher*

Simberloff D (1980) A succession of paradigms in ecology: essentialism to materialism and probabilism. Synthese 43:3–39

Contents

Contributors

F. Bärlocher
Department of Biology
Mount Allison University
Sackville, N.B., EOA 3CO
Canada

A.-C. Chamier
Department of Biology
School of Life Sciences
Royal Holloway
and Bedford New College
Egham, Surrey, TW20 OEX
England

K.R. Chandrashekar
Department of Biosciences
Mangalore University
Mangalagangotri 574 199
Mangalore, India

E.B.G. Jones
School of Biological Sciences
Portsmouth Polytechnic
King Henry I Street
Portsmouth, Hampshire
PO1 2DY
England

K.M. Kaveriappa
Department of Biosciences
Mangalore University
Mangalagangotri 574 199
Mangalore, India

S.T. Moss
School of Biological Sciences
Portsmouth Polytechnic
King Henry I Street
Portsmouth, Hampshire
PO1 2DY
England

S.J. Read
School of Biological Sciences
Portsmouth Polytechnic
King Henry I Street
Portsmouth, Hampshire
PO1 2DY
England

C.A. Shearer
Department of Plant Biology
University of Illinois
289 Morrill Hall
505 South Goodwin Avenue
Urbana, Illinois 61801
USA

K.R. Sridhar
Department of Biosciences
Mangalore University
Mangalagangotri 574 199
Mangalore
India

K. Suberkropp
Department of Biology
University of Alabama
Tuscaloosa, AL 35487
USA

J. Webster
Department
of Biological Sciences
University of Exeter
Exeter, Devon, EX4 4PS
England

1 Research on Aquatic Hyphomycetes: Historical Background and Overview

Felix Bärlocher

1.1 Early Discoveries

Aquatic hyphomycetes are fungi that most commonly occur on dead leaves in streams and rivers and sporulate under water. They are also known as freshwater hyphomycetes (Nilsson 1964), amphibious hyphomycetes (Michaelides and Kendrick 1978), or Ingoldian fungi (Webster and Descals 1981). A majority form tetraradiate conidia (essentially with four diverging arms), some produce sigmoid conidia (long, wormlike, generally curved in more than one plane). Some have spores of more conventional shape.

The first member of this group to be described was *Heliscus lugdunensis* Sacc. et Therry. It was found on pine bark near Lyon, France, and in northern Italy, presumably in a terrestrial environment (Saccardo 1880). It performs the same ecological functions as other aquatic hyphomycetes, but its conidium is clove-shaped rather than tetraradiate or sigmoid.

In the same year, Hartig (1880) described a parasite of maple seedlings as *Cercospora acerina*. Later workers established its identity with *Centrospora acerina*; it is now known as *Mycocentrospora acerina* (Hartig) Deighton (Neergard and Newhall 1951; Deighton 1972). This species is remarkably widespread and versatile; it is a well-known plant pathogen (Wall and Lewis 1980), has been implicated in human infections (Deighton and Mulder 1977) and is also a common stream fungus (Ingold 1975). At least morphologically, there are no consistent differences between the various strains, and Iqbal and Webster (1969) showed that strains they isolated from a stream were pathogenic to carrots and parsnips. It was already known to Hartig (1880) that the fungus grows as a saprophyte as well as a parasite, and he demonstrated that desiccated mycelium can survive for up to one year. This ability is now believed to be important for the dispersal of aquatic hyphomycetes and their continued survival in a given area (Chap. 2).

An important step forward was made by de Wildeman (1893, 1894, 1895). He described four more species of this group, three with tetraradiate and one with sigmoid conidia. He found them in ponds, ditches and marshy areas among algae, on willow leaves and on aquatic macrophytes (e.g. *Hippuris vulgaris* L.). He correctly identified the four-armed structures as conidia. He also noticed the similarity between the spores of *Tetracladium marchalianum* and some old, doubtful and poorly defined genera of algae such as *Asterothrix* and *Cerasteria*. This similarity caused much confusion among phycologists over the next few decades, and many continued to refer to *T. marchalianum* by its algal name (Huber-Pestalozzi 1925, 1938; Brutschy 1927; Klotter 1955). The conidia were often interpreted as representing the entire fungal thallus (of a planktonic fungus), and the two central knobs as spores (e.g. Huber-Pestalozzi 1938). It became known as the "Schmerzenskind" (Huber-Pestalozzi 1938) or "child of sorrow" (Lowe 1927) of the algologists. To complicate matters further, it was often confused with *Tetracladium setigerum* (first described as *Tridentaria setigera*; Grove 1912) or *T. maxilliforme* (described as *Titaea maxilliformis* by Rostrup 1894 from stems of *Trifolium pratense*), and Brutschy (1927) interpreted conidia of *Heliscus lugdunensis* as early stages in the development of *T. marchalianum*. This confusion dates partly back to early drawings by de Wildeman (1893), who apparently did not recognize that he was dealing with more than one *Tetracladium* species. By close observation and spectral analysis of pure cultures Lowe (1927) and Karling (1935) finally proved that the organism was undoubtedly a fungus, though Karling again drew conidia that appear to belong to *T. setigerum*. Nevertheless, it became apparent that the genus *Tetracladium* was fairly widespread in aquatic habitats and occurred on twigs of *Larix*, on rotting *Juncus* leaves and stems, on decaying deciduous leaves, on *Isoetes lacustris*, on *Myriophyllum* (Brutschy 1927; Karling 1935), and in persistent water films in leaf carpets (Scourfield 1940).

Today, few biologists would confuse *T. marchalianum* with *T. setigerum*; nevertheless, the potential for wrong identifications of aquatic hyphomycetes persists and is probably greater than ever. Many newly described species are based on very subtle morphological or developmental differences, some of which can only be observed in pure cultures. Thus, the genus *Tetracladium* now contains seven species, some with overlapping conidial morphology (Roldan et al. 1989). Reports of the occurrence of a given species are therefore always fraught with a degree of uncertainty, especially when identifications are based solely on conidia. Sigmoid spores, in particular, are often very difficult to identify accurately.

In 1906, Kegel discovered *Varicosporium elodeae* on dead or dying shoots of *Elodea canadensis*. He was the first to observe that prolific sporulation can be induced when agar blocks overgrown with the fungus are brought into distilled water. *Tetracladium setigerum* was described from leaves of *Angelica sylvestris* (as *Tridentaria setigera*, Grove 1912); and *Casaresia sphagnorum* on *Sphagnum* by Fragoso (1920). The conidia of the latter are fairly common in streams and are sometimes found attached to leaves. The fungus has been observed growing on wood but not on leaves (Nilsson 1964; Ingold 1975).

1.2 Ingold's Breakthrough

A breakthrough came in 1942, when Ingold discovered the typical habitat of these fungi. At the time, he was looking for chytrids in an alder-lined stream behind his house (C.T. Ingold, 1990, pers. comm.). To his utter amazement he discovered a large collection of fungal spores in scum trapped behind a barrier of twigs. It took him about three months to connect these conidia to mycelia growing and sporulating on alder and willow leaves. When he placed freshly collected leaves in shallow dishes of waters, rich crops of conidiophores appeared within two days. They generally grew at a right angle to the substratum, ensuring that the spore was liberated at a distance from the leaf and was carried away without becoming entangled in the immediate vicinity. Pure cultures on nutrient media usually did not produce conidia until a small strip of agar was submerged in distilled water.

In his first of many papers on aquatic hyphomycetes, Ingold illustrated 16 species, ten of them new. Development of the spores differed greatly among the 16 species but their shapes were of two basic types: tetraradiate or sigmoid. This indicates parallel evolution, and Ingold (1942, 1953) listed three possible functions of the complex spore shape: (1) it might slow down sedimentation and thus ensure adequate dispersal; (2) it might allow the conidium to act as an anchor and become entangled in a suitable substrate; (3) the complex shape might protect the conidia against ingestion by small aquatic animals.

Around the time of Ingold's discovery, a young zoologist associated with the university of London, H.B.N. Hynes, investigated the gut contents of stone fly larvae. He found large quantities of vascular plant remains and frequently small, star-shaped structures. He was unable to identify them and realized only much later that they were conidia of aquatic hyphomycetes (H.B.N. Hynes, in address given at Annual Meeting of North American Benthological Society, Tuscaloosa, Alabama, 1988). He recognized, however, that leaves provide much of the food for stream communities, and he and his associates later established the role of fungi as intermediaries between leaves and invertebrates (Hynes 1963; Kaushik and Hynes 1971). Hynes's observations provided an early, though unpublished, clue that the shape probably does little to defend the spore against ingestion by macroinvertebrates. Indeed, conidia of aquatic hyphomycetes are quite common in the guts of *Gammarus pulex*; some apparently survive passage through the gut (Bärlocher 1981).

1.3 Significance of the Spore Shape

The potential function of the conidial shape was studied experimentally by Webster (1959). He was able to produce large numbers of conidia by forcibly aerating strips of agar overgrown with fungus. His study showed that tetraradiate spores as a group did not have consistently lower sedimentation rates than spores of other shapes. Compared to current flow in a stream (typically around $100\,cm\,s^{-1}$, Hynes 1970), spore sedimentation was very slow (around $0.1\,mm\,s^{-1}$) and probably insignificant in the turbulent conditions prevalent in streams. The aquatic yeast, *Vanrija aquatica*, may be an exception. It was first isolated from foam in a mountain tarn (Jones and Sloof 1965), and later discovered in hollow internodes of dead *Equisetum* shoots. The shape of its conidia is affected by nutrient availability (Webster and Davey 1975). Conidia from starving cultures develop long, tapering arms, and sediment more slowly than those from well-nourished cultures with short, fat arms. Webster and Davey (1975) suggested that the fungus is sedentary when growing in contact with decaying plant tissues; when nutrients are scarce, it becomes planktonic and is dispersed by water currents.

When spores were passed by a glassrod coated with collodion, the results were unequivocal: the ten most efficiently captured species all had tetraradiate spores, and the next three sigmoid spores (Webster 1959). The trapping of sigmoid spores increased when they were mixed with tetraradiate spores, probably due to the roughening of the surface.

Webster (1959) thus demonstrated that the function of the spore shape of aquatic hyphomycetes is probably to minimize downstream transport. When a tetraradiate spore makes contact with a surface it does so at three points and acts as a tripod, which represents a very stable form of attachment. Germination requires a contact stimulus and upon settling, spore arms expand to form a pad or appressorium, which further strengthens adhesion to surfaces (Webster 1959). An extensive investigation of spore settling and germination was recently completed by Read (1990).

Sigmoid spores in a low speed current tend to bump along the bottom, and conidial ends make frequent contact with the surface (Webster and Davey 1984). When they come to rest, they usually have two points of contact: at one end and at a point along the length. They quickly germinate (within 1–2 h); generally, an appressorium is formed at one end, and the conidium swings around so that it is aligned parallel to the direction of the current. The functional significance of spore shapes in freshwater and marine fungi was recently reviewed by Webster (1987). Spore settling and germination are further discussed in Chapter 7.

An alternative explanation of the spore shape was offered by Bandoni (1975) in connection with the terrestrial occurrence of many of these fungi. The tetraradiate shape might facilitate the dispersal in aqueous films between layers of terrestrial leaf litter.

Finally, the mere fact that conidia of aquatic hyphomycetes are comparatively large increases their probability of encountering a target (Cox 1983). In a random search, probability of contact will increase with the effective width of the search vehicle, and the diameter of aquatic conidia is approximately 10–40 times as long as that of the average terrestrial spore.

Air bubbles rising through water efficiently capture fungal spores, and examination of foam is believed to give a reasonably complete list of aquatic hyphomycetes occurring in a given stream (Ingold 1975). Not all spores are trapped equally well, however, and those with complex branching tend to be overrepresented (Iqbal and Webster 1973a). Sigmoid spores, on the other hand, are relatively scarce (Lindsey and Glover 1976).

1.4 Advances in Taxonomy

Ingold's discovery (1942) ushered in an era where reports of the occurrence of these fungi multiplied. It has been called the starting point of a "minor mycological industry" (Ainsworth 1976). Among the more comprehensive studies are those dealing with California (Ranzoni 1953), Japan (Tubaki 1957), the eastern United States Petersen 1962, 1963a,b), Scandinavia (Nilsson 1964), and the Soviet Union (Dudka 1985). The great taxonomic advances that have been made is well illustrated by Nilsson's remark (1964) that the number of species found in Sweden, 39, was higher than any reported from other countries. Today, a comparable or higher number is often routinely identified from a single stream. In 1981, Webster and Descals listed over 150 known species, today there are close to 300 (L. Marvanová 1991, pers. comm.). Ranzoni (1956) discovered the first sexual state of an aquatic hyphomycete; today there are over 20 such connections (Chap. 5).

Ingold continued to describe new species from all over the world (complete bibliography in Dick et al. 1985). His *Guide to Aquatic Hyphomycetes* (Ingold 1975) summarized taxonomic knowledge to that point, and contains much useful information on occurrence, ecology and techniques to study these fungi. A comprehensive, up-to-date monograph of aquatic hyphomycetes is being prepared by E. Descals, L. Marvanová and J. Webster.

1.5 Substrata Colonized by Aquatic Hyphomycetes

Aquatic hyphomycetes occur on almost any type of plant debris and also superficially on glass and plastic. Wood is of particular importance because of its long residence time, and because it is the main site of sexual reproduction (Chaps. 4 and 5). Most commonly, dead branches and twigs serve as substrata; however, aquatic hyphomycetes also occur on or in roots (Waid 1954). They appear to be quite common in roots of riparian alder trees (Fisher et al. 1991).

Macrophytes sometimes replace autumn-shed leaves as the major source of detrital material in streams; the first and so far only thorough mycological study of these materials was conducted by Kirby (1984); some of his results were published in Kirby et al. (1990). He compared fungal colonization of alder leaves with that of *Ranunculus penicillatus* (which can contribute up to 70% of total particulate input in many eastern England streams), *R. flammula, Nasturtium officinale, Apium nodiforum*, and *Potamogeton polygonifolis*. Fungal diversity on macrophytes was clearly lower: *Tetracladium marchalianum* and *Pythium* sp. (Oomycetes) were generally the dominant species. Overall, only nine other aquatic hyphomycetes were found during the entire study. By comparison, it is not uncommon to find 8–12 species on a single alder leaf (Ingold 1975). Fungal biomass per unit weight of aquatic macrophyte was estimated to be an average of 0.8% (maximum: 1.5%).

Thus, it appears that aquatic hyphomycetes are indeed most common on deciduous leaves, in particular those of *Alnus*, *Fraxinus* and *Quercus* sp., and beginning with Triska (1970) and Newton (1971), many scientists have followed colonization of various leaves throughout the seasons (Chap. 3). However, contrary to earlier beliefs (e.g. Ranzoni 1953), they also colonize conifer twigs and needles. As mentioned earlier, the first member of the group, *Heliscus lugdunensis*, was discovered on pine bark (Saccardo 1880), and Brutschy (1927) found the same species as well as *Tetracladium marchalianum* on *Larix* twigs. More recently, Arnold (1970) discovered several species on needles and twigs of *Pseudotsuga* sp. They were later reported from needles of *Abies, Larix, Picea, Pinus, Sequoia, Taxus* and *Thuja* (Bärlocher and Oertli 1978a,b; Bärlocher et al. 1978, 1979; Michaelides and Kendrick 1978; Iqbal et al. 1980; Summerbell and Cannings 1981; Bärlocher 1982; Sati et al. 1989). The relatively slow fungal colonization of conifer needles appears to be due to antifungal substances which are kept in the needle for considerable periods of time by well developed epidermal structures.

Latex of rubber leaves allowed growth and sporulation of five species of aquatic fungi (one belonging to the aero-aquatic group). Only one fungus was able to grow, but not sporulate, on latex of papaya leaves. Experiments with calotropis, banyan, and jack gave intermediate results (Sridhar and Kaveriappa 1987).

As mentioned earlier, *Mycocentrospora acerina*, common in streams, is also a fairly widespread plant pathogen. Other aquatic hyphomycetes have since been implicated in plant diseases (Watanabe 1975; Webster and Descals 1981). More surprisingly, *Pleuropedium tricladioides* has been found growing in a freshwater nematode (Nolan 1977), and *Anguillospora pseudolongissima* apparently hyperparasitizes *Glomus epigaeus* and *G. fasciculatus* (Daniels and Menge 1980). Recently, an aquatic Basidiomycete with tetraradiate conidia (*Camptobasidium hydrophilum*) has been discovered that appears to function primarily as an unspecialized, biotrophic parasite of aquatic hyphomycetes (Marvanová and Suberkropp 1990).

1.6 Geographical Distribution and Temperature Preferences

Aquatic hyphomycetes are cosmopolitan in their distribution, extending from the arctic circle to the equator (Kobayasi et al. 1967, 1971; Müller-Haeckel and Marvanová 1976, 1979; Webster and Descals 1981; Engblom et al. 1986; Bhat and Chien 1990). *Flagellospora curvula*, *Lemmoniera aquatica*, *L. centrosphaera*, and *Alatospora acuminata* are some of the dominant species in the high north. In tropical latitudes, species such as *Lunulospora curvula*, *Triscelophorus acuminatus* and *Ingoldiella hamata* are more typical (Chap. 10).

Optimal temperatures for in vitro growth and sporulation are broadly correlated with the geographical occurrence of the fungi. In those of temperate regions, vegetative growth is generally highest between 15–25 °C, whereas sporulation often reaches a peak at somewhat lower temperatures (Tubaki 1957; Thornton 1963; Price and Talbot 1966; Koske and Duncan 1974; Danninger et al. 1979; Suberkropp and Klug 1981; Suberkropp 1984). The maximum temperature where growth occurs is often quite low, sometimes as low as 25–30 °C. At the other extreme, Canadian aquatic hyphomycetes could be isolated from plates incubated at 0 °C, and caused weight loss of maple leaves at the same temperature (Bärlocher and Kendrick 1974). Agar cultures survived at −20 °C for one month, and viable spores were found in frozen foam, and on snow above the tree limit in Sweden (Nilsson 1964). Many of the common aquatic hyphomycetes can survive several cycles of freezing and thawing without appreciable loss of viability (J. Webster 1990, pers. comm.).

In five tropical species, maximum growth was around 25 °C, and was much lower at 30 °C (Singh and Musa 1977).

It is important to remember, however, that stream temperature usually declines with altitude. Consequently, in mountain streams of tropical regions, species normally more common in high latitude regions will occur (see Chap. 10). Conversely, seasonal increases of temperature will favour the periodic appearance of "tropical" species in temperate streams. Thus, *Lunulospora curvula* is fairly abundant from summer to early autumn in Devon streams (Iqbal and Webster 1973b). Its subsequent decline may be due to its inability to grow at low temperatures (Suberkropp 1984; Webster et al. 1976). The same correlation, extended to *Flagellospora penicillioides* and *Clavatospora tentacula*, was found in a Michigan stream (Suberkropp 1984).

1.7 Nutrition

Aquatic hyphomycetes generally have no absolute requirement for vitamins, though an external supply may shorten the lag phase (Thornton 1963). Only *Volucrispora aurantiaca* Haskins (probably not a true aquatic hyphomycete) showed an absolute requirement for biotin. A majority can use nitrate and ammonium as nitrogen sources (again with the exception of *V. aurantiaca* Haskins), and a wide range of monosaccharides, oligosaccharides, and amino acids (Thornton 1963, 1965).

There are some early reports that aquatic hyphomycetes can degrade cellulose and other plant polymers (Nilsson 1964; Thornton 1963; Tubaki 1957). The first thorough study was conducted by Suberkropp and Klug (1980). Five species were grown with hickory leaf material as sole source of carbon and energy; all led to skeletonization of leaves through maceration, indicating attack by pectin-degrading enzymes. Leaf cells were released from the leaf, suggesting a previously unknown fungal role in the production of fine particulate organic matter, which provides the food for filter-feeders (Wallace and Merritt 1980). Leaves skeletonized by fungal activity are recognizable by numerous small holes resembling pin-pricks, and are easily distinguished from those due to activity of shredders, which ingest most of the finer veins.

All five species studied by Suberkropp and Klug (1980) showed pH dependent activity toward cellulose, xylan, and pectin, and this has been confirmed for most species tested later (Danninger et al. 1979; Chamier and Dixon 1982, 1983; Suberkropp et al. 1983; Chamier et al. 1984; Zemek et al. 1985; Chandrashekar and Kaveriappa 1988; Abdullah and Taj-Aldeen 1989; Chauvet 1989). Ca ions were reported to stimulate pectin lyase activity and growth (Chamier and Dixon 1983).

Little work has been done on ligninases, partly because there are no simple assays comparable to those used for polysaccharides. There is at least circumstantial evidence that some species can attack lignin. Thus, Leightley and Eaton (1977) observed cavity and erosion decay by *H. lugdunensis*, and Jones (1981) reports that four out of six species caused soft rot attack (but not *H. lugdunensis*); Fisher et al. (1983) used two lignin-agar preparations; out of 16 species, four reacted positively with one, four positive with the other, and two with both preparations. Zemek et al. (1985) tested 23 species for degradation of natural lignin derived from *Nicotiana tabacum* L., and all showed some breakdown. The work done on enzymes of aquatic hyphomycetes was reviewed by Chamier (1985) and Suberkropp (1991b) and is further discussed in Chapters 4, 6 and 8.

There is some evidence that fungal growth in streams is influenced by inorganic nutrients, particularly by nitrogen and phosphorus compounds (Suberkropp 1991a).

1.8 Sporulation

Ingold (1975) defined aquatic hyphomycetes as those species that regularly sporulate under water. Kegel (1906) was the first to show that sporulation can be induced when an agar block overgrown with *Varicosporium* is submerged. With nutrient-rich agar media, though not always with plain agar, submersion is often an absolute requirement (Nilsson 1964).

Webster (1959) showed that sporulation can be enhanced by forcibly aerating chopped agar cultures in water using a stream of compressed air. This is unlikely to be related to oxygen concentration, however, and there is no evidence for volatile sporulation inhibitors (Webster and Towfik 1972); it is therefore probably due to mechanical effects of increasing turbulence.

In a later study, Webster (1975) related increased sporulation to two phenomena: the time for spore development is much shorter at high aeration rate, and the number of conidiophores per unit area of culture surface is greater. However, there is no simple linear correlation between water flow and sporulation (Sanders and Webster 1980). Of 17 species, eight sporulated almost as prolifically at low flow rates as at more rapid rates; nine species showed a lower flow rate limit below which sporulation was significantly reduced. The members of the first group are fairly common in terrestrial as well as aquatic habitats, while those of the second group are rare outside streams.

Sporulation also appears to be stimulated by low organic nutrient content, as well as by low mineral content (Aimer 1989). Nilsson (1964) reported that all his colonies sporulated freely on plain agar, but not on agar with added nutrients. I also found this to be true for most cultures isolated during my Ph.D. work (unpubl. obs.); other workers could not confirm these observations (Aimer 1989). The pH value, as well as buffer concentrations, can selectively inhibit or stimulate sporulation of some species (Iqbal 1976).

Aquatic hyphomycetes can often be identified by their spores; based on this, Iqbal and Webster (1973b) introduced a simple and elegant technique to characterize fungal communities of streams. A known volume of water is sucked through a membrane filter (pore size 5–8 μm); the spores are trapped by the filter, where they can be stained, counted and often identified. This allowed, for the first time, the quantification of the extreme seasonal fluctuations in the activity of the fungi (Chap. 3). One must be aware, however, that some spores captured in this way may have been washed in from the soil, or may even have dripped in from overhanging trees (Ando and Tubaki 1984; Tubaki et al. 1985). Another potential drawback is that unequivocal identification, based exclusively on conidia, is not always possible.

The membrane filtration technique has also been used to estimate the reproductive potential of mycelia present on leaves at various stages of decay (Bärlocher 1981, 1982). A leaf, or leaf disk, is suspendend in 2–500 ml distilled, sterilized water and aerated for 1–2 days. The supernatant is then filtered and the spores counted and identified. If the leaf material is dried and weighed, spore production per weight can be calculated. In two days, up to 8 spores per μg leaf material (close to 1 million per leaf) can be produced. Nilsson (1964), who did similar studies without aeration, achieved maxima of over 300 000 in 24 h.

When sporulation occurs in the absence of nutrients, mycelium with cell contents declines, and empty hyphae accumulate, indicating that spore production "uses up" vegetative mycelium (Aimer 1989). When grown under standard conditions on agar media, a set proportion of mycelial biomass of a particular species will be converted into a set number of spores. Therefore the length or volume of mycelium present in a leaf can potentially be calculated from the number of spores produced by a fungus. At maximum sporulation rates, $1\,mm^3$ spore volume per day required from 3 to $22 \times 10^6\,mm^3$ mycelium in six species; in two species (*Alatospora pulchella* and *Anguillospora longissima*), the values were much higher. The conversion from mycelium to spores suggested by these ratios seems extremely low, and, in view of the high production of spores from stream-collected leaves (Nilsson 1964; Bärlocher 1982), must be much higher under natural conditions. Findlay and Arsuffi (1989) estimated that the carbon released as spores compared to the production of hyphal carbon ranged from 18–122%. Suberkropp (1991a) demonstrated that growth and sporulation are closely correlated in aquatic hyphomycetes.

1.9 Role in Food Webs

By the late 1960s, research into taxonomy, biology and ecology of aquatic hyphomycetes was well advanced. Nevertheless, few stream ecologists were probably aware of these fungi. This changed when their involvement in the food web was discovered. Two studies demonstrated that several stream invertebrates prefer to eat conditioned leaves (leaves colonized by fungi) rather than sterile leaves (Triska 1970; Kaushik and Hynes 1971). The importance of leaves for stream communities, and the mediating role of fungi, are further discussed in Chapter 6. Ironically, many of these early experiments were done under conditions where aquatic fungi are unlikely to have been present.

These discoveries increased the need for techniques to measure fungal biomass on leaves, which has always been one of the more difficult problems of mycology. In early studies leaves were cleared and the mycelium stained (Iversen 1973; Bärlocher and Kendrick 1974); later, ATP was used as an index of fungal biomass (Suberkropp et al. 1976; Rosset et al. 1982); Findlay and Arsuffi (1989) used an epifluorescence technique. The relative contribution of fungi and bacteria was often estimated by using selective antibiotics (e.g. Triska 1970; Kaushik and Haynes 1971; Aimer 1989) or by light (Iversen 1973) or epifluorescence microscopy (Findlay and Arsuffi 1989).

1.10 Water Chemistry

It seems self-evident that the water chemistry will influence the occurrence and activity of aquatic hyphomycetes. Suzuki and Nimura (1961) looked at the distribution of the fungi in relation to pH, and concluded that they were more common in "harmonic" lakes (pH 6.2–6.5) rather than acidotrophic lakes (pH 3.8–5.8). Lakes, of course, are not the primary habitat of aquatic hyphomycetes. Iqbal (1976) investigated the effect of pH on the sporulation by selected species, and Bärlocher and Rosset (1981) tried to establish correlations between occurrence of fungi in a stream and its water chemistry. This topic is further discussed in Chapter 8. The range of habitats apparently suitable for aquatic hyphomycetes was extended when Jones and Oliver (1964) found them in brackish water (Chap. 2). Inevitably, the effects of acid precipitation, of heavy metals and pesticides on activity of aquatic hyphomycetes have been studied in recent years. They are discussed in Chapters 8 and 9.

References

Abdullah SK, Taj-Aldeen SJ (1989) Extracellular enzymatic activity of aquatic and aero-aquatic conidial fungi. Hydrobiologia 174:217–223

Aimer RD (1989) Ecology of aquatic hyphomycetes in New Zealand streams. Thesis, University of Waikato, New Zealand

Ainsworth GC (1976) Introduction to the history of mycology. University of Cambridge Press, Cambridge

Ando K, Tubaki K (1984) Some undescribed hyphomycetes in the rain drops from intact leaf-surface. Trans Mycol Soc Jpn 25:21–37

Arnold GRW (1970) Aquatische Hyphomyzeten auf Koniferen. Limnologica 7:381–382

Bandoni RJ (1975) Significance of the tetraradiate form in dispersal of terrestrial fungi. Rep Tottori Mycol Inst 12:105–113

Bärlocher F (1981) Fungi on the food and in the faeces of *Gammarus pulex*. Trans Br Mycol Soc 76:160–165

Bärlocher F (1982) Conidium production from leaves and needles in four streams. Can J Bot 60:1487–1494

Bärlocher F, Kendrick B (1974) Dynamics of the fungal population on leaves in a stream. J Ecol 62:761–791

Bärlocher F, Oertli JJ (1978a) Colonization of conifer needles by aquatic hyphomycetes. Can J Bot 56:57–62

Bärlocher F, Oertli JJ (1978b) Inhibitors of aquatic hyphomycetes in dead conifer needles. Mycologia 70:964–974

Bärlocher F, Rosset J (1981) Aquatic hyphomycete spora in two Black Forest and two Swiss Jura streams. Trans Br Mycol Soc 76:351–355

Bärlocher F, Kendrick B, Michaelides J (1978) Colonization and conditioning of *Pinus resinosa* needles by aquatic hyphomycetes. Arch Hydrobiol 81:462–474

Bärlocher F, Oertli JJ, Guggenheim R (1979) Accelerated loss of antifungal inhibitors from *Pinus leucodermis* needles. Trans Br Mycol Soc 72:277–281

Bhat DJ, Chien C-Y (1990) Water-borne hyphomycetes found in Ethiopia. Trans Mycol Soc Jpn 31:147–158

Brutschy A (1927) *Tetracladium marchalianum* de Wildeman, ein nur wenig beobachteter Wasserpilz. Mikrokosmos 28:145–146

Chamier A-C (1985) Cell-wall degrading enzymes of aquatic hyphomycetes: a review. Bot J Linn Soc 91:67–81

Chamier A-C, Dixon PA (1982) Pectinases in leaf degradation by aquatic hyphomycetes: the enzymes and leaf maceration. J Gen Microbiol 128:2469–2483

Chamier A-C, Dixon PA (1983) Effect of calcium ion concentration on leaf maceration by *Tetrachaetum elegans*. Trans Br Mycol Soc 81:415–418

Chamier A-C, Dixon PA, Archer SA (1984) The spatial distribution of fungi on decomposing alder leaves in a freshwater stream. Oecologia 64:92–103
Chandrashekar KR, Kaveriappa KM (1988) Production of extracellular enzymes by aquatic hyphomycetes. Folia Microbiol 33:55–58
Chauvet E (1989) Production, flux et décomposition des litières en milieu alluvial. Dynamique et rôle des hyphomycètes aquatiques dans le processus de décomposition. Thesis, Paul Sabatier University, Toulouse
Cox PA (1983) Search theory, random motion and the convergent evolution of pollen and spore morphology in aquatic plants. Am Nat 121:9–31
Daniels BA, Menge JA (1980) Hyperparasitism of vesicular-arbuscular mycorrhizal fungi. Phytopathology 70:584–588
Danninger E, Messner K, Rohr M (1979) Untersuchungen über den biologischen Abbau organischer Naturstoffe durch aquatische Hyphomyzeten. Zentralbl Bakteriol Hyg 1 Abt Orig B 169:282–286
Deighton FC (1972) *Mycocentrospora*, a new name for *Centrospora* Neerg. Taxon 21: 716
Deighton FC, Mulder JL (1977) *Mycocentrospora acerina* as a human pathogen. Trans Br Mycol Soc 69:326–327
De Wildeman E (1893) Notes mycologiques. Fasc II Ann Soc Belge Microscopie 17: 35–68
De Wildeman E (1894) Notes mycologiques. Fasc III Ann Soc Belge Microscopie 18: 135–161
De Wildeman E (1895) Notes mycologiques. Fasc II Ann Soc Belge Microscopie 19: 193–206
Dick MW, Pegler DN, Sutton BC (1985) Contributions to mycology. Acad Press, New York
Dudka IA (1985) Aquatic Fungi imperfecti from the Soviet Union. Ukrainian Academy of Science, Kiev
Engblom E, Lingdell P-E, Marvanová L, Müller-Haeckel A (1986) Foam spora in running waters of southern Greenland. Polar Res 4:47–51
Findlay SEG, Arsuffi TL (1989) Microbial growth and detritus transformations during decomposition of leaf litter in a stream. Freshwater Biol 21:261–269
Fisher PJ, Davey RA, Webster J (1983) Degradation of lignin by aquatic and aeroaquatic hyphomycetes. Trans Br Mycol Soc 80:166–168
Fisher J, Webster J, Petrini O (1991) Aquatic hyphomycetes and other fungi in living aquatic and terrestrial roots of *Alnus glutinosa*. Mycol Res 95:543–547
Fragoso GR (1920) Nuevo genero y especie de hifal sobre hojas de *Sphagnum*. Bol Soc Esp Hist Nat 20:112–114
Grove WB (1912) New or noteworthy fungi – Part IV. J Bot (London) 50:9–18
Hartig R (1880) Der Ahornkeimlingspilz, *Cercospora acerina*. Untersuch Forstbotan Inst, München, 1:58–61
Huber-Pestalozzi G (1925) Zur Morphologie und Entwicklungsgeschichte von *Asterothrix (Cerasterias) raphidioides* (Rheinsch) Printz. Hedwigia 65:169–178
Huber-Pestalozzi G (1938) Das Phytoplankton des Süsswassers. Schweizerbart'sche Verlagsbuchhandlung, Stuttgart
Hynes HBN (1963) Imported organic matter and secondary productivity in streams. Proc XVI Int Congr Zool 4:324–329
Hynes HBN (1970) The ecology of running waters. University of Toronto Press, Toronto
Ingold CT (1942) Aquatic hyphomycetes of decaying alder leaves. Trans Br Mycol Soc 25:339–417
Ingold CT (1953) Dispersal in fungi. Clarendon Press, Oxford
Ingold CT (1975) An illustrated guide to aquatic and waterborne hyphomycetes (Fungi Imperfecti) with notes on their biology. Freshwater Biol Assoc Sci Publ No 30
Iqbal SH (1976) Effect of pH on sporulation of freshwater hyphomycetes. Biologia (Lahore) 22:143–153
Iqbal SH, Webster J (1969) Pathogenicity of aquatic isolates of *Centrospora acerina* to carrots and parsnips. Trans Br Mycol Soc 53:486–490

Iqbal SH, Webster J (1973a) The trapping of aquatic hyphomycete spores by air bubbles. Trans Br Mycol Soc 60:37–48

Iqbal SH, Webster J (1973b) Aquatic hyphomycete spora of the River Exe and its tributaries. Trans Br Mycol Soc 61:331–346

Iqbal SH, Bhatty SF, Malik KS (1980) Freshwater hyphomycetes on submerged decaying pine needles in Pakistan. Trans Mycol Soc Jpn 21:321–327

Iversen TM (1973) Decomposition of autumn-shed beech leaves in a spring-brook and its significance for the fauna. Arch Hydrobiol 73:305–312

Jones EBG (1981) Observations on the ecology of lignicolous aquatic hyphomycetes. In: Wicklow DT, Carroll GC (eds) The fungal community. Marcel Dekker, New York, pp 731–742

Jones EBG, Oliver AC (1964) Occurrence of aquatic hyphomycetes on wood submerged in fresh and brackish water. Trans Br Mycol Soc 47:45–48

Jones EBG, Sloof W (1965) *Candida aquatica* sp nov isolated from water scums. Antonie Leeuwenhoek J Microbiol Serol 32:223–228

Karling JS (1935) *Tetracladium marchalianum* and its relation to *Asterothrix*, *Phycastrum*, and *Cerasterias*. Mycologia 27:478–495

Kaushik NK, Hynes HBN (1971) The fate of the dead leaves that fall into streams. Arch Hydrobiol 68:465–515

Kegel W (1906) *Varicosporium elodeae*, ein Wasserpilz mit auffallender Konidienbildung. Ber Dtsch Bot Ges 68:465–515

Kirby JMH (1984) Microbial aspects of aquatic macrophyte decomposition. Thesis, University of Exeter, Exeter

Kirby JJH, Webster J, Baker JH (1990) A particle plating method for analysis of fungal community composition and structure. Mycol Res 94:621–626

Klotter HE (1955) Über Wasserpilze (mit besonderer Berücksichtigung von *Lemonniera aquatica*). Mikrokosmos 44:125–128

Kobayasi Y, Hiratsuka N, Korf RP, Tubaki K, Aoshima K, Soneda M, Sugiyama J (1967) Mycological studies of the Alaskan Arctic. Annu Rep Inst Ferment Osaka 3:1–138

Kobayasi Y, Hiratsuka N, Otani Y, Tubaki K Udagawa S-I, Sugiyama J, Konno K (1971) Mycological studies of the Angmagssalik region of Greenland. Bull Natl Sci Mus (Tokyo) 14:1–96

Koske RE, Duncan IW (1974) Temperature effects on growth, sporulation and germination of some "aquatic" hyphomycetes. Can J Bot 52:1387–1391

Leightley LE, Eaton RA (1977) Mechanism of decay of timber by aquatic microorganisms. Brit Wood Pres Assoc, Annu Conv, pp 1–26

Lindsey BI, Glover BJ (1976) Ecological studies of spores of aquatic hyphomycetes in the Cringle Brook, Lincs. Hydrobiologia 51:201–208

Lowe CW (1927) *Cerasterias*, the child of sorrow of the algologist. Trans R Soc Can III, 21:C111

Marvanová L, Suberkropp K (1990) *Camptobasidium hydrophilum* and its anamorph, *Crucella subtilis*: a new heterobasidiomycete from streams. Mycologia 82:208–217

Michaelides J, Kendrick B (1978) An investigation of factors retarding colonization of conifer needles by amphibious hyphomycetes in streams. Mycologia 70:419–430

Müller-Haeckel A, Marvanová L (1976) Konidienproduktion und -kolonisation von Süsswasser-Hyphomyzeten im Kaltisjokk (Lappland). Bot Not 129:405–409

Müller-Haeckel A, Marvanová L (1979) Periodicity of aquatic hyphomycetes in the subarctic. Trans Br Mycol Soc 73:109–116

Neergard P, Newhall AG (1951) Notes on the physiology and pathogenicity of *Centrospora acerina* (Hartig) Newhall. Phytopathology 41:1021–1033

Newton JA (1971) A mycological study of decay in the leaves of deciduous tress on the bed of a river. Thesis, University of Salford, England

Nilsson S (1964) Freshwater hyphomycetes: taxonomy, morphology and ecology. Symb Bot Ups 18:1–130

Nolan RA (1977) Physiological studies with the aquatic hyphomycete *Pleuropedium tricladioides* isolated from the freshwater nematode *Neomesomermis flumenalis*. Mycologia 69:914–926

Petersen RH (1962) Aquatic hyphomycetes from North America. I Aleuriosporae (Part 1), and key to the genera. Mycologia 54:117–151

Petersen RH (1963a) Aquatic hyphomycetes from North America. II Aleuriosporae (Part 2) and Blastosporae. Mycologia 55:18–29

Petersen RH (1963b) Aquatic hyphomycetes from North America. III Phialosporae and miscellaneous species. Mycologia 55:570–581

Price IP, Talbot PHB (1966) An aquatic hyphomycete in a lignicolous habitat. Aust J Bot 14:19–23

Ranzoni FV (1953) The aquatic hyphomycetes of California. Farlowia 4:353–398

Ranzoni FV (1956) The perfect stage of *Flagellospora penicillioides*. Am J Bot 43:13–17

Read SJ (1990) Spore attachment in fungi with special reference to freshwater hyphomycetes. Thesis, Portsmouth Polytechnic, Portsmouth

Roldan A, Descals E, Honrubia M (1989) Pure culture studies on *Tetracladium*. Mycol Res 93:452–465

Rosset J, Bärlocher F, Oertil JJ (1982) Decomposition of conifer needles and deciduous leaves in two Black Forest and two Swiss Jura streams. Int Rev Gesages Hydrobiol 67:695–711

Rostrop E (1894) Mykologiske Meddelelser IV. Bot Tidsskr 19:36–47

Saccardo PA (1880) Conspectus generum fungorum Italiae inferorium. Michelia 2:1–38

Sanders PF, Webster J (1980) Sporulation responses of some "aquatic hyphomycetes" in flowing water. Trans Br Mycol Soc 74:601–605

Sati SC, Mer GS, Tiwari N (1989) Occurrence of water-borne conidial fungi on *Pinus roxburghii* needles. Curr Sci 58:918

Scourfield DJ (1940) The microscopic life of the "leaf carpet" of woods and forests. Essex Nat 26:231–246

Singh N, Musa TM (1977) Terrestrial occurrence and the effect of temperature on growth, sporulation and spore germination of some tropical aquatic hyphomycetes. Trans Br Mycol Soc 68:103–106

Sridhar KR, Kaveriappa KM (1987) Culturing water-borne hyphomycetes on plant latex. J Ind Bot Soc 66:232–233

Suberkropp K (1984) Effect of temperature on seasonal occurrence of aquatic hyphomycetes. Trans Br Mycol Soc 82:53–62

Suberkropp K (1991a) Relationships between growth and sporulation of aquatic hyphomycetes on decomposing leaf litter. Mycol Res 95:843–850

Suberkropp K (1991b) Aquatic hyphomycete communities. In: Carroll GC, Wicklow DT (eds) The fungal community. Marcel Dekker, New York (in press)

Suberkropp K, Klug MJ (1980) The maceration of deciduous leaf litter by aquatic hyphomycetes. Can J Bot 58:1025–1031

Suberkropp K, Klug MJ (1981) Degradation of leaf litter by aquatic hyphomycetes. In: Wicklow DT, Carroll GC (eds) The fungal community. Marcel Dekker, New York pp 761–776

Suberkropp K, Godshalk GL, Klug MJ (1976) Changes in the chemical composition of leaves during processing in a woodland stream. Ecology 57:720–727

Suberkropp K, Arsuffi TL, Anderson JP (1983) Comparison of degradative ability, enzymatic activity and palatability of aquatic hyphomycetes grown on leaf litter. Appl Environ Microbiol 46:237–244

Summerbell R, Cannings RA (1981) Fungi and *Brillia retifinis* (Diptera: Chironomidae) in the decomposition of conifer leaves in a British Columbia stream. Syesis 14:125–130

Suzuki S, Nimura N (1961) Relation between the distribution of aquatic Hyphomycetes in Japanese lake types. Bot Mag Tokyo 74:51–55

Thornton DR (1963) The physiology and nutrition of some aquatic hyphomycetes. J Gen Microbiol 33:23–31

Thornton DR (1965) Amino acid analysis of fresh leaf litter and the nitrogen nutrition of some aquatic hyphomycetes. Can J Microbiol 11:657–662

Triska FJ (1970) Seasonal distribution of aquatic hyphomycetes in relation to the disappearance of leaf litter from a woodland stream. Thesis, University of Pittsburgh, Pittsburgh

Tubaki K (1957) Studies on the Japanese hyphomycetes III Aquatic group. Bull Natl Sci Mus (Tokyo) 41:249–268

Tubaki K, Tokumasu S, Ando K (1985) Morning dew and *Tripospermum* (Hyphomycetes). Bot J Linn Soc 91:45–50

Waid JS (1954) Occurrence of aquatic hyphomycetes upon the root surfaces of beech grown in woodland soils. Trans Br Mycol Soc 37:420–421

Wall CJ, Lewis BG (1980) Infection of carrot plants by *Mycocentrospora acerina*. Trans Br Mycol Soc 74:587–593

Wallace JB, Merritt RW (1980) Filter-feeding ecology of aquatic insects. Ann Rev Entomol 25:103–132

Watanabe T (1975) *Tetracladium setigerum*, an aquatic hyphomycete associated with gentian and strawberry roots. Trans Mycol Soc Jpn 16:348–350

Webster J (1959) Experiments with spores of aquatic hyphomycetes. I Sedimentation, and impaction on smooth surfaces. Ann Bot 23:595–611

Webster J (1975) Further studies of sporulation of aquatic hyphomycetes in relation to aeration. Trans Br Mycol Soc 64:119–127

Webster J (1987) Convergent evolution and the functional significance of spore shape in aquatic and semi-aquatic fungi. In: Rayner ADM, Brasier CM, Moore D (eds) Evolutionary biology of the fungi. Cambridge University Press, Cambridge, pp 191–201

Webster J, Davey RA (1975) Sedimentation rates and trapping efficiency of cells of *Candida aquatica*. Trans Br Mycol Soc 64:437–440

Webster J, Davey RA (1984) Sigmoid conidial shape in aquatic fungi. Trans Br Mycol Soc 83:43–52

Webster J, Descals E (1981) Morphology, distribution and ecology of conidial fungi in freshwater habitats. In: Cole GT, Kendrick B (eds) Biology of Conidial Fungi, vol 1. Acad Press, New York, pp 295–355

Webster J, Towfik FH (1972) Sporulation of aquatic hyphomycetes in relation to aeration. Trans Br Mycol Soc 59:353–364

Webster J, Moran ST, Davey RA (1976) Growth and sporulation of *Tricladium chaetocladium* and *Lunulospora curvula* in relation to temperature. Trans Br Mycol Soc 67:491–549

Zemek J, Marvanová L, Kuniak L, Kadlecikova B (1985) Hydrolytic enzymes in aquatic hyphomycetes. Folia Microbiol (Prague) 30:363–372

2 Recent Developments in Stream Ecology and Their Relevance to Aquatic Mycology

Felix Bärlocher

2.1 Longitudinal Variation: Distinct Zones or Gradual Replacement?

If one follows a stream from its source to the sea, many of its physical and geomorphological properties, such as current speed, temperature, depth, etc., change in a predictable manner (Hynes 1970). All these factors influence occurrence and abundance of the various stream organisms. It comes as no surprise therefore that running waters display a longitudinal biological zonation. Fish biologists were the first to use this phenomenon to subdivide the sections of a watercourse into four zones named after their dominant fish. This empirical approach was developed in the 1950s in Belgium and northern Germany and proved useful for fishery biologists in nearby areas. However, presence or absence of some of these species in a given stream or river often depends on historical accidents and, with increasing distance from the region where the concept was first proposed, the original four species were replaced by others.

The same approach can be, and in many cases has been, used for any other sufficiently well-known group of organisms. They all show more or less predictable changes in numbers and kinds of species along a water course. If, at certain points, more species disappear or appear than at others, these points could serve as boundaries of natural zones, each characterized by its own unique assemblage of species. This interpretation has been championed by Illies and Botosaneanu (Illies 1961; Illies and Botosaneanu 1963; Botosaneanu 1979). Based on the distribution of fishes and invertebrates, they suggested that rivers can be subdivided into three major zones, with relatively distinct boundaries: crenon (spring), rhithron (stream; relatively cold stenothermic fauna, high O_2 concentration, fast and turbulent flow, bed of rocks, stones or gravel, few sandy or silty patches) and potamon (river; monthly mean temperature rises to over 20 °C, O_2 deficits may occur, flow slower, often laminar, bed mainly mud and sand). The different regions are inhabited by distinct biocenoses (a community of organisms that are mutually dependent and act as a unit, Trepl 1987). Hynes (1970) characterized these terms as convenient shorthand descriptions; useful, as long as they are not taken as articles of faith. Based on a comprehensive review of the literature he concluded, however, that zones are rarely if ever discrete, and that rivers are clines in an ecological sense. Transition, rather than zonation, is dominant. The gradual, continuous nature of changes along a water course has been accepted by most stream ecologists today. Some dissenting voices remain. Botosaneanu (1979) remained convinced that the fauna does not change as gradually as abiotic factors. He believes that there are three distinct biocenoses or ecologically similar clusters of species. Geographically distant streams with similar physics and chemistry will have strikingly parallel series of species with the same adaptations ("Lebensformtypen"). A similar philosophy forms the basis of Cummins' (1973, 1974) use of functional groups, rather than actual species, to characterize invertebrate communities. It allows combining genetically diverse entities into groups that share fundamental properties. His system is based on morphological-behavioural differences of food acquisition. He differs from Botosaneanu in assuming that transitions in the dominance of the various groups within a river system are gradual.

The emphasis on functional groups or Lebensformtypen makes obvious sense, if one wants to compare geographically widely separated streams. They may share only a very few species, yet the basic processes and structure of their communities may be similar.

As one of the advantages of a classification based on zones, Botosaneanu (1979) mentions its "elasticity", or capability of incorporating new findings. A somewhat similar stance was taken by Pennak (1971). He dismisses the idea of treating the whole stream or river as a single ecological entity. According to him, only stream segments, stream sections, or stream habitats may be properly compared and contrasted from place to place. Each should be described by 13 parameters with 4–6 categories each. This defines over 180 million types, which, as Hynes (1975) put it ". . . is really little more useful than listing them by name". But this may be appropriate since ". . . every stream is likely to be an individual and thus not really very easily classifiable" (Hynes 1975).

More recently, Statzner and Higler (1986) concluded that stream hydraulics (physical characteristics of stream flow) is the single most important factor governing the zonation of stream benthos. It has been known for a long time that microdistribution of invertebrates is often related to hydraulic parameters (Hynes 1970). Statzner and Higler (1986) claim that stream hydraulics often exhibit distinct changes localized within relatively short stretches, and that these changes are reflected by the composition of stream communities. The areas, where discontinuities occur, therefore represent natural boundaries.

How does all this apply to aquatic hyphomycetes? To begin with, as a group they are more restricted in their nutritional habits than invertebrates. Almost all of them grow best on leaves and branches of deciduous and conifer trees, and are much less common on grasses and aquatic macrophytes (Chaps. 1, 4, 5). The majority perform basically the same function, which is the breakdown of structural plant polysaccharides. Some species are more versatile or more active than others (Suberkropp and Klug 1980, Chamier 1985, Suberkropp 1991); at present, there are insufficient data to allow further subdivision of their functional roles.

Because fungi depend on allochthonous litter, riparian vegetation will be a decisive factor in determining the fungal community in streams. In addition, water chemistry probably plays a role (Chap. 8). Nothing is known about the effects of stream hydraulics on the occurrence and abundance of fungi. Through its effects on the retention of leaves and other debris and more directly on the settling of spores, it will affect them at least indirectly by determining whether leaves are retained or swept downstream, and whether spores are able to settle and colonize new substrata. Working with *Lycopodium* spores in model streams, Reynolds et al. (1990) concluded that settling loss closely follows an exponential decay function. Its slope (settling vs. time) decreases with water depth but seems largely unaffected by water velocity. The velocity nevertheless influences the distances travelled by suspended particles. As a consequence, downstream transport is most pronounced in deep, fast-flowing streams.

In areas of excessive silt or mud deposition, leaves often turn black due to lack of oxygen, and show very little evidence of fungal colonization. There seems little doubt that a typical potamon section (river, with a silty or muddy bed) is not very favourable for aquatic hyphomycetes. Nilsson (1964) reported that they are absent from very large rivers. In the Garonne (80 m wide, 2 m deep), Chauvet (1989) found only 16 species on willow leaves during 18 weeks. Aquatic hyphomycetes were absent in "dead arms". In slow flowing Illinois rivers, the summer mycota consists of 3–4 species, in the fall its rises to 10–12 (C.A. Shearer 1990, pers. comm.).

There are two major studies of longitudinal variation in aquatic hyphomycetes. Shearer and Webster (1985) studied three sites along the River Teign (England). The major difference was found between sites I and II, separated by 4 km. They had two distinctly different communities. The lowest number of species was found at site I, an acid moorland stream where *Juncus* culms were the dominant substratum. Differences between sites II and III (separated by 12 km, and both in deciduous forests) were much less clear-cut, and most species occurred at II and III.

Gönczöl (1975) studied the Morgó stream (Mts. Börzsöny, Hungary). Leaf input in the upper reaches was dominated by *Fagus sylvatica* and *Carpinus betulus*; in the lower reaches, these trees were replaced by *Alnus glutinosa*. Gönczöl observed an abrupt change in the communities of aquatic hyphomycetes: "two well distinguishable alliances of species can be recognized, species which intermingle not or only to a small extent: one in the upper reaches, the others in the lower reaches" [sic].

In a later study, Gönczöl (1989) tried to correlate species occurrences with water hardness. He concluded that both water hardness and substrate influence the structure of the fungal community.

From these studies, it appears that the rate of community change along a water course can vary considerably. Relatively abrupt changes in invertebrate communities may be due to changes in stream hydraulics; with fungi, riparian vegetation and water chemistry are probably the primary determinants, though hydraulics, by determining retention of particulate matter, may well modify their effects (see Sect. 2.6). The physical concepts of flow that are relevant to organisms were discussed by Vogel (1981).

2.2 The River Continuum Concept

The gradual, continuous nature of changes along a water course forms the backbone of the River Continuum concept (Vannote et al. 1980; expanded and updated in Cummins et al. 1984; Minshall et al. 1985) which integrates biological features of flowing waters with physical and geomorphic gradients. What does it offer to the fungal ecologist? The starting point is sound: it proposes that understanding biological strategies and dynamics of river systems requires consideration of the gradient of physical factors formed by drainage networks. Insofar as the community is regulated by fluvial geomorphic processes, "this gradient should elicit a series of responses within the constituent populations resulting in a continuum of biotic adjustments and consistent patterns of loading, transport, utilization, and storage of organic matter along the length of a river". As mentioned earlier, this has been disputed by Statzner and Higler (1985): in their view, patterns of physical parameters do not form a "continuous" or "intergrading" gradient. Again, part of the disagreement may be purely semantic. But even if we accept gradual changes in the physical and chemical parameter, this does not guarantee a similar pattern in the biological community, since there is always the possibility of non-linear relationships, leading to chaotic patterns (Chap. 3).

According to the RCC (as introduced by Vannote et al. 1980), biological communities in natural streams assume strategies involving minimum energy loss. This has been criticized by Winterbourn et al. (1981) as uncomfortably close to the idea of group or even community selection. It is equivalent to saying that leakage of organic material in running waters (downstream transport from one section to the next) will be minimized due to the biological community. This is not necessarily true: downstream transport was actually less when the invertebrate fauna was destroyed in experiments or in computer simulations (Wallace et al. 1982; Webster 1983). Similarly, fungal activity accelerates transformation of coarse particulate organic matter into fine particles, and therefore presumably downstream transport (Suberkropp and Klug 1980), unless this is compensated for by higher filter-feeding activity. Vannote et al. (1980) probably meant to say that a community, shaped by a network of interactions, is more efficient at preventing leakage than a single species. This is often true: the wastes of one species (e.g. shredder faeces) are generally used by others (e.g. collectors), leading to food chains and webs. There is no obvious reason, however, why this should lead to a straightforward connection between the diversity or evolutionary age of a community and its efficiency. More species, or longer periods of time, could mean more, or new, potential feedbacks, and there is always the possibility that any one of them profoundly changes the outcome. The net effect cannot, as system analysts like to say, be predicted by "hand waving".

An important fact in stream ecology is the unidirectional movement of water. This means, among many other things, that anything suspended in the water (ions, molecules, organic particles, organisms) will be transported downstream for a certain distance before resettling. This phenomenon is referred to as "spiralling" (Mulholland et al. 1985). It practically ensures a certain blurring of possible boundaries: downstream sections will inevitably be influenced by upstream sections. The spiralling distance (distance travelled by ions, molecules or particles between their release from the stream bed and their resettlement) appears to be determined primarily by the type of spiralling substance, current velocity, and presence and effectiveness of physical retention devices (Minshall et al. 1985). Obviously, this process is of great significance for fungi. It affects their dispersal and the availability of their substrates. It is discussed further in Section 2.6.1.

As pointed out by Hynes (1963), valleys are usually humid areas. Stream and river banks are therefore often densely covered with vegetation. Where conditions permit, shrubs and trees dominate. Their canopies may completely cover small streams. Remains of the terrestrial vegetation, such as autumn-shed leaves, conifer needles, twigs, etc., fall into the water and serve as food primarily for invertebrates. The contribution of this allochthonous material has recently been reviewed by Minshall (1988), Webster and Benfield (1986) and Winterbourn (1986).

As we go downstream, the channel widens and more light reaches the water. This allows more plant growth in the water and the amount of terrestrial plant detritus added per volume water declines. In other words, the food supply shifts from allochthonous to autochthonous sources. In large rivers, light may again limit photosynthesis in the water due to increasing depth and turbidity. In areas with slow-moving water, phytoplankton may become important; more commonly, organic substances imported from upstream or smaller tributaries become the primary source of food. These changes in the spectrum of food items should influence the composition of the consumer community: leaf-shredders should dominate in the upper reaches and be replaced by collectors in the lower reaches. In any reach, influx of organic pollutants, another potential source of food, can distort this progression.

These predictions of the RCC corresponded reasonably well to the conditions in rivers of four distinct geographic areas in the USA (Minshall et al. 1983). In Britain, several predictions of the RCC were confirmed, but may have been due to factors independent of the RCC (Townsend and Hildrew 1984). Of course, the vegetation and landscape of Britain have been extensively modified over the centuries, and a good match cannot reasonably be expected. But confirmation was also unconvincing or absent in relatively pristine New Zealand streams (Winterbourn et al. 1981; Ryder and Scott 1988), due to the unpredictability of rainfall patterns, poor retention of suspended materials by stream beds, and unstructured, opportunistic invertebrate communities. When reviewing some aspects of the RCC, Townsend (1989) made the following comment "... the idealized downstream changing pattern in primary trophic resource ... is remarkable primarily because it is not usually realized and cannot provide a world-wide generalization."

Changes in the supply of allochthonous litter will obviously influence fungal abundance. Other factors being equal, the amount of suitable leaf material per volume water and its residence time in a given reach are probably crucial. If they decrease, one would expect a drop in fungal activity. This could occur by widening or deepening of rivers; by changes in riparian vegetation (type or quantity); by changes in leaf-shredder activity; or, by changes in the retentive properties of the stream bed. The low retention of leaf litter in many New Zealand streams is thought to be responsible for the relative scarcity of shredders; on the other hand, a survey of aquatic hyphomycetes over three years yielded some 43 species (Aimer and Segedin 1985). Given the uncertainties of correct identifications, and the geographic isolation of New Zealand, these values fall within the range reported from other geographic areas.

Undoubtedly, there will be a decline in activity of aquatic hyphomycetes as we move downstream; there is no compelling reason why this decline should be gradual or abrupt. Local conditions will be decisive.

As one of the recent advances in stream ecosystem theory, Minshall et al. (1985) mention "the progression from an individualistic to a holistic viewpoint" with a major emphasis on process-oriented studies and a functional view of the various groups of organisms. Mycologists and bacteriologists have long used a process-oriented and functional approach, often more out of necessity than out of choice. It means that much of their effort has been concerned with estimating biomass and its turnover, and enzymatic activities. These are undoubtedly important measurements, but like all process-oriented, holistic approaches, they offer only limited insights into mechanisms that determine which, and how many, species can share the same function in a given stream. The holistic viewpoint does not concern itself with how competition and predation might influence absence or presence of species. It tends to reduce contributions of microbial ecologists to dutiful, bland statements about the significance of microbes in the great scheme of things.

Likewise, the emphasis on the entire stream system rather than separate reaches may be important to provide a general framework for comparing streams. But every mycologist knows that fungal communities change over very small distances. Within each zone, it is possible to distinguish fairly definite fungal communities associated with particular types of substrata; fungal occurrence is based on presence in discrete patches, and highly seasonal (Chap. 3). As one passes downstream, individual types of patches increase or decrease in area; if the inherent variability is glossed over, trends become recognizable. The advice given by Minshall et al. (1985) "... in examining conditions along a river system, samples should be taken so that the mean or most characteristic geomorphic conditions in each stretch are represented" sounds rather dubious to fungal ecologists. The patch dynamics concept, which emphasizes temporal phenomena and focuses on the importance of history and chance, seems more appropriate (Pringle et al. 1988; Townsend 1989). It emphasizes that every section is patchy on some scale. The basic subunits are discrete micromosaics, which nevertheless participate in a very active exchange among themselves.

Based on practical experience, mycologists know that they are most likely to find aquatic hyphomycetes in the upper reaches of streams, and undoubtedly will continue to concentrate their efforts there. They will continue to be most interested in similar sections of different streams and longitudinal studies within the upper reaches of stream. There is some numerical justification: headwaters are more common than any other stream section.

I suspect Wilson's (1971) remarks about the decline of the superorganism concept in social insects capture the way many fungal ecologists feel about holistic studies: "... it exemplifies the way inspirational, holistic ideas in biology often give rise to experimental, reductionist approaches that supplant them. For the present generation, which is so devoted to the reductionist philosophy, the superorganism concept provided a very appealing mirage. It drew us to a point on the horizon. But, as we worked close, the mirage dissolved – for the moment at least – leaving us in the midst of unfamiliar terrain, the exploration of which came to demand our undivided attention."

In a defence of the RCC, Minshall et al. (1985) show that the RCC yields nothing to the zone concept as far as "elasticity" is concerned. It can incorporate almost any newly appreciated exceptions, and is highly retentive of new insights. But many of the concepts or tenets usually associated with it are not really integral; they do not contradict it, but neither are they essential. The RCC represents a loose grouping of interesting developments in ways we can look at streams. Thus, Statzner and Higler (1985) believe that five tenets generally associated with the RCC are in conflict with the current state of knowledge or are at least open to various interpretations. Even though among the five questionable tenets is what they consider the central statement of the RCC that "biological communities should become established which approach equilibrium with the dynamic physical conditions of the channel" (Vannote et al. 1980), they do not as a consequence reject the RCC.

Many of the publications discussing the RCC are interesting and should be read by mycologists working with stream fungi, but is the RCC essential to their work? I believe the answer is (to borrow a phrase by Tudge, 1989): "up to a point"; which is Evelyn Waugh's euphemism for "not really".

2.3 Hyporheos and Groundwater

So far nothing has been said about the origin of water making up streams and rivers. Overland flow is rare in most cases (Freeze and Cherry 1979; Hynes 1983). This means that most of the water that enters a stream has been in contact with the soil for some time. Groundwater is water that has percolated down through an unsaturated zone to the water table; interflow infiltrates soil but does not move down to the water table. Instead it moves laterally toward the stream channel. The specific path that water takes, the geology of the valley and its terrestrial vegetation largely determine types and amounts of inorganic ions dissolved in streams (Likens 1984). Output of these ions also varies appreciably between headwaters, where the soil is often thin and porous and acid cations predominate, and lowland rivers where hydrologic flow paths are much more complex, and soil retention times and channel lengths greater. These inorganic ions will influence many microbially controlled processes, among them fungal or bacterial decomposition of organic substances (Chap. 8).

Maybe more significant, however, is the fact that interflow and groundwater also carry dissolved organic substances (DOM), in other words, potential food (Hynes 1983). These are normally adsorbed and metabolized at the surfaces of stones, pebbles and small particles making up the stream sediment. Such interfaces are covered with a biofilm, called epilithon, consisting of various microorganisms embedded in polysaccharides. The epilithon is the primary site for removal of DOM carried by open stream water, and many invertebrates use this layer as food (Lock 1981; Madsen 1972; Meyer et al. 1988; Rounick and Winterbourn 1983; Winterbourn 1986).

There is no reason why the same processes should not be involved when interflow or groundwater with its load of dissolved organic matter percolate upward through the stream sediment. This represents yet another source of energy for stream communities. Its magnitude is unknown. Still, as pointed out by Hynes (1983), if we are to get a realistic idea of how streams function, it is vital that we learn about all potential sources of energy. Considering the fact that the majority of stream animals, the so-called hyporheos, often live and presumably feed at a considerable distance underneath the sediment/water boundary (Williams 1981, 1989), it may well turn out that it is through this hidden entrance that the community receives much of its energy. If this proves to be the case, the RCC with its emphasis on visible, above-ground food sources (leaves and stream plants) is obviously to "superficial" to give an accurate description of stream systems, and H.B.N. Hynes may qualify as the Sigmund Freud of stream ecology.

2.4 Pristine Rivers and Their Boundaries

Triska (1984) and Sedell and Froggatt (1984) showed that even rivers which we consider to be close to their natural state have been extensively modified by human intervention. Groundwater tables used to be higher. As a result much of the riparian forest was at least occasionally flooded, creating a variety of standing or slowly moving water bodies. The river itself was often subdivided into several main channels, separated by frequently shifting sandbanks or islands. The overall effect was that even large rivers were more intimately connected with, and more strongly influenced by, their terrestrial surroundings, and presumably provided more habitats where aquatic hyphomycetes could flourish. In the example studied by Sedell and Froggatt (1984), >250 km of the river edge originally were in contact with riparian forest; by 1967, this had decreased to 64 km. In the same period, the ratio edge/river length had decreased from >10.0 to 2.6. Conversely, North American beavers were nearly extinct 90 years ago; they have since multiplied substantially. This has significantly increased water surface area and associated boundaries with uplands (Naiman et al. 1988).

Riparian vegetation is more than a supplier of energy for stream communities. Tree roots help stabilize banks and thus lower erosion; increased load of particles carried in the water may alternately smother or sandblast stream organisms. Large woody debris dams can profoundly change water flow patterns. Their higher frequency of occurrence in undisturbed streams makes these more retentive by reducing erosion. Especially in larger rivers, most of these "snags" have been removed to facilitate navigation and prevent flooding of the surrounding areas.

In summer, small streams running through cleared land often have their water temperature raised by a few degrees compared to shaded streams. This seemingly negligible difference is largely held responsible for the presently very restricted distribution of the brook trout in southern Ontario and the Appalachians (Hynes 1970). In both regions, much of the riparian vegetation has been removed. Fungi may be similarly affected. Provided enough substrata remain, increased temperature may favour species such as *Lunulospora curvula* and *Flagellospora penicillioides* (Webster et al. 1976; Suberkropp 1984; Chap. 3).

Boundaries between streams and surrounding terrestrial environment have been compared to semipermeable membranes (Naiman et al. 1988). They modify direction, character and magnitude of materials exchanged by adjacent ecological systems. Boundaries are of obvious relevance to aquatic hyphomycetes, not only because this is where their food comes from, but also as potential refuge and reservoir for recolonization (Sect. 2.6.). Current changes, such as the one near Toulouse, France, where a pioneer forest is being replaced by an oak hardwood forest, and thereby changes from a system influenced by seasonal flooding to a predominantly terrestrial system no longer heavily influenced by flooding (Naiman et al. 1988), are particularly suitable to study the influence of the boundary on fungal stream communities. Also of great interest would be a study of latitudinal gradients in boundary characteristics, shifting from tundra to moist tropical forests as we approach the equator.

2.5 The Effects of Altitude and Latitude

Most stream ecologists have studied woodland streams in temperate zones. A more balanced view would include moorland streams, temporary streams, prairie streams, blackwater streams, as well as streams in other climatic zones (Hynes 1970; Matthews 1988; Meyer 1990; Minshall 1988; see Chap. 10). At high altitudes and in high latitudes, streams never warm far above freezing point. There is, nevertheless, a definite, although limited very-cold-water fauna. It is also found in mountain streams in the tropics, for example in the permanently cold tundra zone on Mount Kenya (Hynes 1970). Very little is known about fungi in such streams, except from the studies of a subarctic stream by Müller-Haeckel and Marvanová (1976, 1979a). While temperature is unlikely to be a limiting factor (Chap. 3), the scarcity or absence of suitable substrata in streams above the tree line may be decisive.

In tropical streams of low altitudes, animal species richness is not uniformly higher than in comparable communities of temperate streams (Covich 1988). Invertebrates have not been widely categorized in the neotropics; they appear to be dominated by generalized consumers.

Inorganic nutrients are generally, but not always, very dilute in tropical streams. On the other hand, the generally-held assumption that tropical forests take up and retain higher concentrations of nutrients in their biomass than temperate forests is not supported by recent investigations (Covich 1988).

There are very marked regional differences in the timing of leaf fall. It is affected by rainfall patterns, episodic storm events and types of forest species present. Some species lose a small proportion of their leaves throughout the year, others lose them as the dry season approaches, a third group drops them after the onset of the rainy season.

Because of the longer growing season, some tropical streams receive nearly twice the amount of terrestrial detritus as temperate streams. In some rainforests, much of the shredding of leaves is done by terrestrial invertebrates during the dry season. In others, leaves accumulate on the ground and begin to break down only when the rainy season allows renewed fungal growth. Many different fungal groups may be involved in leaf breakdown. Thus, in a Puerto Rico rainforest, terrestrial Basidiomycetes reduced downhill movements of leaves on steeply sloped stream banks by 40% (Lodge and Asbury 1988). Aerial rhizomorphs of *Marasmius*, *Psathurella*, and others function much like spider webs and entrap leaves before they reach the ground (Covich 1988; Hedger 1990). Palms and tree ferns often have fronds or large leaves that are retained on the tree after death. They sometimes form a basket-like structure that catches leaves and other detritus from nearby trees. Many of these "containers" hold water, and are colonized by detritivores. Little is known about the fungi present in these habitats; in Hungary, aquatic hyphomycetes were present in water-filled tree holes (Gönczöl 1976).

It appears, then, that the path of leaves from tree to stream is often much less straightforward in the tropics. In many cases, when leaves finally enter streams they may have already be extensively modified by aerial and terrestrial fungi. The effects on aquatic hyphomycetes are unknown.

2.6 Drift and Compensatory Mechanisms

Within streams, passive movements are predominantly unidirectional. There are, of course, zones of dead water and reversed flow but their occurrence and effects are restricted (Hynes 1970). One of the perennial preoccupations of stream ecologists has been to find convincing mechanisms that allow organisms with little or no active movements to persist in a given stream reach. If water current were the only effective mechanism, their gradual depletion and eventual elimination from streams would seem inevitable. The key to their continued presence in streams must therefore lie in processes that reverse the effects of streamflow.

2.6.1 Transport Within and Between Streams

Sampling procedures which use stream water will give information on fungi in transport. Fungi may be present as independent particles (spores, pieces of mycelium) or attached to substrata.

Organic particles in streams are probably never sterile. The distances they travel are therefore relevant for microbial movements. For aquatic hyphomycetes, leaves, twigs and branches are the primary substrata. Of these, the fates of leaves and particles derived from them are known best. In temperate regions, 1 m^2 of stream bed area may receive 4–600 g leaves per year (Bird and Kaushik 1981). These leaves already carry terrestrial fungi and yeasts. These may contribute to leaf decomposition; in general, however, they decline steadily in the stream and are replaced by aquatic hyphomycetes, often preceded by a short phase of Peronosporales and Saprolegniales (Chap. 3). In one study, the average distance travelled by leaves falling into a stream was 200 m; within 1000 m all leaves were generally entrained (Young et al. 1978). In two South African streams, the distances from the release point necessary to retain 90% of the leaves were 19 and 91 m (Prochazka et al. 1991). The retentiveness of a stream reach is strongly influenced by current speed, channel morphology and riparian vegetation; leaves are captured more quickly when current speed is low, channel structure heterogeneous and density of shrubs high (Speaker et al. 1988). Prochazka et al. (1991) concluded that discharge played a major role in determining the retentiveness of their streams. After the initial downstream displacement, transport of large particles is often insignificant, except during spates (Dance 1981; Malmquist et al. 1978). Even then, retention can be high: more than a third of artificial leaves (plastic strips) were retained within 660 m during high flows in Oregon streams (Speaker et al. 1988).

Most suspended organic particles are very small (for example, only 10–20% were larger than 106 μm in a study by Sedell et al. 1978). They may have been formed by mechanical or enzymatic maceration of leaves (Suberkropp and Klug 1980), they may be faeces, precipitates of dissolved organic matter, or "slime" sloughed off from rocks in the stream bed (Lock 1981). Very little is known about types and quantities of microorganisms they carry. It is known that they are harvested by filter-feeding invertebrates, and it is assumed that the microbial biomass is digested preferentially (Ward and Cummins 1979). The distances travelled before a particle is trapped generally increase with stream size. It seems, then, that the likelihood and magnitude of downstream transport are negatively correlated with the size and weight of an object and positively with stream order.

The same downstream movements affect detached conidia and pieces of mycelium. Conidia are usually small (normally the diameter is 50–100 μm). As a consequence, they offer little resistance to the current, and they are also liable to being ingested by filter-feeding invertebrates. They can reach concentrations of up to 30 000 per litre (Webster and Descals 1981). It is unknown how many of them will successfully colonize a new substratum and how many will die and be incorporated into the food web. Gönczöl (1975, 1976) estimated that, in a headwater stream, they will be carried downstream for no more than three to four km before being removed from the water. If, as has been shown for other small particles (Reynolds et al. 1990), their removal from the water column follows a negative exponential model, the loss coefficient would have to be $-0.0015\,m^{-1}l^{-1}$ ($-0.0023\,m^{-1}l^{-1}$) to lower the number of suspended spores to 1% (0.1%) of its original value within 3000 m. At these loss rates, and if we further assume that the release of newly formed spores (in $m^{-1}l^{-1}$) is constant, 31 (47) spores would have to be added to reach a final concentration of 30 000. Half-lives (characterized by stream length that would remove 50% of the spores from the water) would be 453 m (301 m), or, if we use 4 km as standard, 603 m (401 m). These rough estimates may not be far out of line; Suberkropp et al. (1988) found a maximum of 149 spores per litre near the source of a stream.

A more elaborate model based on firmer empirical evidence was recently published by Thomas et al. (1991a,b). In their stream, the half-lives of three species (all with tetraradiate conidia) varied between 690 and 810 m. In three other species (two with sigmoid, one with tetraradiate conidia), spore numbers were added in a geometric fashion. This suggests an autocatalyic pattern: spore production downstream may be influenced by spores imported from upstream. Or, gradual changes in physicochemical parameters influenced spore production in a non-linear fashion. A third group did not show any consistent patterns of spore dynamics, and could therefore be considered a "variable input" group. Thomas et al. (1991a,b) make the interesting suggestion that these different patterns may reflect different fungal strategies for exploiting resources. The "variable input" species may be more specific in their requirements than those with continuous input, whereas "arithmetic input" species may colonize a wide range of substrata, or specialize on continuously available substrata such as wood.

Spores are units of dispersal. This process may be advantageous for two reasons (Howe and Smallwood 1982): (1) It may protect the offspring from disproportionately high mortality near the parent (escape hypothesis); (2) it may allow the colonization of uncompetitive environments as they open up (colonization hypothesis). Both possibilities are relevant for aquatic hyphomycetes. By colonizing leaves, they render them more palatable for stream invertebrates, thereby endangering their food supply as well as their lives. In addition, leaves rapidly become less suitable for growth and reproduction (Bärlocher 1982). These conditions will favour species that spread very quickly from substrate to substrate after leaf fall. Since conidia are non-motile, their dispersal depends on water movements. Not surprisingly, sporulation is stimulated by water flow (Webster 1975; Sanders and Webster 1980). Inevitably, the conidia will be carried downstream with the risk of overshooting their target and being swept out of the more favourable upper reaches of a stream. Their size and shape are obvious attempts to minimize this danger (Chaps. 1, 7). While water current is indispensable for the dispersal of aquatic hyphomycetes, their most striking adaptations seem designed to lower its effectiveness.

Mechanisms that might compensate for downstream displacement have been discussed by several authors (Nilsson 1964; Ingold 1976; Webster and Descals 1979). One possibility is transport on or in animals. In the terrestrial environment beetle-associated mites are important vectors of ascospores (Moser et al. 1989). In streams, single spores, or foam, which can accumulate enormous numbers of spores, or small pieces of leaves colonized by the fungi might become attached to animals (aquatic invertebrates, emerging insect imagines, water fowl, etc.) and be transported to other areas. To my knowledge, no data confirming or disproving this possibility have as yet been published. It has been shown, however, that many fungi can survive ingestion by animals. For example, conidia or conidium-forming structures of eight aquatic hyphomycetes survived passage through the digestive tract of *Gammarus pulex* (Bärlocher 1981). There was a considerable reduction in viability (conidium production from leaves: $300\,mg^{-1}$; faeces: $30\,mg^{-1}$).

Nevertheless, this clearly establishes the potential for compensatory movements. Freshwater Gammaridae are generally efficient upstream migrants (Hynes 1970; Williams 1981, 1989). Rossi and Vitagliano-Tadini (1978) demonstrated that fungi survived on faeces of *Asellus* and were able to recolonize sterile leaves. Finally, Thornton (1970, 1971) showed that many protoctistan fungi (Chytridiales, Peronosporales, Saprolegniales) can survive ingestion by earthworms and several bird species. Ingestion by the latter might allow long distance dispersal between different watersheds.

Nilsson (1964) observed conidia of aquatic hyphomycetes up to 30 m away from waterfalls. Gas bubbles bursting at the water-air interface often eject considerable amounts of organic matter and microorganisms into the air (Blanchard and Parker 1977), and Iqbal and Webster (1973) found that microscope slides held above aerated spore suspensions were soon "splattered by spore-containing droplets". Foam, as mentioned earlier, often contains high numbers of conidia, and therefore acts as an excellent retention device. It is often blown onto leaves lying at the edge of streams. Again, it is unknown over what distances wind-borne propagules can be distributed and remain viable. Asexual spores of aquatic fungi are generally too fragile to make long-range dispersal likely. Sexual spores are more promising candidates. At least 20 aquatic hyphomycetes are known to have a sexual state (Webster and Descals 1981, Chap. 5). The small sexual spores may well be transported over very long distances which might help explain the presence of similar fungal floras in the same climatic regions of different continents (Ingold 1975; Ranzoni 1979). Their involvement in the annual cycle in streams remains entirely unknown.

In recent years, it has become increasingly clear that many stream communities extend vertically into the so-called hyporheal zone (Hynes 1983; Williams 1981). This habitat often supports high populations of invertebrates, and provides protection against spates. It might have a similar function for microorganisms. In one study, 21% of the estimated leaf input became buried, and the weight of buried detritus (0–20 cm deep) exceeded that of surface detritus by an order of magnitude (Metzler and Smock 1990). Stream beds are often surprisingly well oxygenated (Williams 1981, 1989) and would therefore allow fungal respiration. In addition, some aquatic hyphomycetes can survive for limited periods of time under anaerobic conditions (Field and Webster 1983).

The basal flow of most running waters is derived from groundwater, and discharge areas (where groundwater enters the stream) may alternate with recharge areas (where water flows downward from the stream) along the stream channel (Hynes 1983). Thus, there is an exchange between groundwater and streamwater, and fungi may be transported in both directions. A study of an artesian well in central Texas revealed 35 taxa, dominated by terrestrial hyphomycetes and Zygomycetes (Kuehn and Koehn 1988). No aquatic hyphomycetes were found, but saprolegniaceous species were reasonably common. Glass-beads buried 10 cm deep in streams occasionally trapped tetraradiate and sigmoid conidia, indicating that spores can penetrate quite deeply into the sediment (Bärlocher and Murdoch 1989).

2.6.2 Exchanges with the Surrounding Terrestrial Habitats

As mentioned above, many aquatic hyphomycetes have a sexual state. Generally, it is found on moist but not submerged wood. Some "aquatic" fungi may actually be amphibious with a terrestrial sexual state and an aquatic asexual state (Chap. 5). This would, of course, immediately solve the problem of the fungi's persistence in streams: the terrestrial component, which is not subject to unidirectional displacement, could provide a stationary source of new inoculum.

There is convincing evidence that, even without a predominantly or exclusively terrestrial sexual state, many aquatic hyphomycetes can survive in soils (e.g. Waid 1954; Gönczöl 1976; Singh and Musa 1977; Goos 1978; Gönczöl and Révay 1983; Sridhar and Kaveriappa 1987; Fisher and Petrini 1989). Park (1974) and Bandoni (1981) list a total of 34 "aquatic" species that have also been found in soil (for comparison, Webster and Descals listed 165 Ingoldian species in 1981). Of the 34, six are aero-aquatic fungi, and 28 could be interpreted as aquatic hyphomycetes (based on spore morphology). For at least three of them, the main habitat is undoubtedly terrestrial, since they have never been observed elsewhere. This demonstrates the danger of equating tetraradiate spore morphology with stream fungus. The same shape is also common in marine fungi (Kohlmeyer and Kohlmeyer 1979) and in terrestrial Coelomycetes (Müller and Loeffler 1982).

Roots of riparian trees are another potentially important habitat for aquatic hyphomycetes. To date, a total of 12 species have been isolated from alder roots (Fisher et al. 1991). Gönczöl (1976) also reported aquatic hyphomycetes from tree hollows.

Some species appear to be equally common in terrestrial and aquatic environments. For example, *Varicosporium elodeae* was isolated from root surfaces of beech grown in woodland soil (Waid 1954), successfully colonized various cellulosic substrates in soil (Park 1982), and is often common on leaves in streams (Webster and Descals 1981). Similarly, *Mycocentrospora acerina* is a well-known plant pathogen (Wall and Lewis 1980a), has been implicated as a human pathogen (Deighton and Mulder 1977) and is also a common stream fungus (Ingold 1975). Its conidia can survive up to 20 weeks in soil, its chlamydospores up to two years (Wall and Lewis 1978, 1980b). At least morphologically, there are no consistent differences between the various strains.

Most aquatic hyphomycetes, however, occur more abundantly and consistently in streams than anywhere else. This has been demonstrated most convincingly by Webster (1977). He collected leaf samples at 10 m intervals along a steeply sloping bank. At higher levels, conidia and number of species declined drastically, and even at the lowest sites, which were subject to occasional floods, they were much less common than on stream leaves.

Still, even limited survival on land may be important for providing sources for reinoculation of emptied stream reaches, or temporarily dried-up streams (Ranzoni 1953). Floods remove litter from the banks, concentrate and bury it at other sites, but they also deposit stream leaves with their fungi on the banks (Bell and Sipp 1975; Mayack et al. 1989). In such leaves, fungi survived for up to one year (Thakur 1977; Sanders and Webster 1978; Sridhar and Kaveriappa 1988b), though survival was much reduced when leaves were tied to twigs of riparian trees (Sanders and Webster 1978). The tetraradiate spores may spread quickly over wet litter layers with the help of monolayers covering water films (Bandoni and Koske 1974).

2.6.3 Exchanges with the Sea

Most streams eventually empty into the sea whose high salt content presents a new challenge to freshwater organisms. Regular changes between low and high tide may retain newly introduced particles and their microorganisms in a transitional zone with brackish water. This raises at least the possibility that some fungi may escape and be transported back into more hospitable regions by the same transport processes active between land and freshwater.

Transport of coarse particulate organic material into estuaries can be considerable. Thus, annual exports from the River Ängerån (10–20 m wide) add up to 500–1000 kg (Müller-Haeckel 1982). Depending on season, this consists predominantly of decaying river plants, algae, or newly shed leaves from riparian trees. It settles within 100–300 m below the river mouth. In addition, up to 50 tons of fine organic material ($<0.6\,\mu m$) leaves the river per year. This material spreads up to 1 km into the estuary. Up to 2×10^{13} conidia of aquatic fungi may leave the river mouth per month (Müller-Haeckel and Marvanová 1982). This would correspond to the number of spores released by at least 200 kg leaves (Bärlocher 1982), equal to the yearly input to 3–500 m^2 of stream. Conidia very seldom travel more than 1500 m into the estuary.

Other systems exporting huge amounts of organic material, together with any microorganisms it may carry, include mangrove swamps and tidal salt marshes (Kohlmeyer and Kohlmeyer 1979; Newell 1976).

Very little is known about the fate of the fungi entering the sea. The occurrence of aquatic hyphomycetes in brackish or sea water has been reported many times (Kirk 1969; Shearer and Crane 1971; Shearer 1972; Müller-Haeckel and Marvanová 1979a,b). Byrne and Jones (1975) and Jones and Byrne (1976) found that reproduction (two species) was severely inhibited by high salt concentrations. Good growth but very little sporulation (eight species) in sea water was also reported by Müller-Haeckel and Marvanová (1979a). In the Nethravathi estuary (India), the highest numbers of species were found when the salinity ranged between 0.5–10.5‰ (Sridhar and Kaveriappa 1988a,b). Growth and spore germination were again less affected than reproduction by increasing salinity.

Woodblocks were colonized by aquatic hyphomycetes in fresh and brackish but not in sea water (Jones and Oliver 1964). Similarly, their colonization of leaves in the Bothnian Sea generally decreased with higher salinity (Müller-Haeckel and Marvanová 1982). Leaves carrying these fungi were almost totally consumed within 14 days. Thus, aquatic hyphomycetes may remain metabolically active in brackish water. Their reproduction and the persistence of their substrata, however, will be much reduced.

It appears, then, that the sea is generally the final stage in the journey of most fungi transported in running waters. They may be important as a food source for the local animals, but it is unlikely that they or their offspring will ever again encounter fresh water.

2.7 Conclusions

One of the major developments in the study of aquatic hyphomycetes has been the realization that their occurrence is not restricted to the stream itself. They have therefore been called "amphibious fungi" (Michaelides and Kendrick 1978; the same term, however, was suggested earlier for fungi growing on halophytes in tidal salt marshes; Pugh 1974).

This parallels similar developments in stream ecology, where it has been realized that streams and their biota extend into the stream bed and beyond the stream banks. The following words, with which Hynes (1975) concluded the Edgardo Baldi lecture at the 19th International Limnological Congress in Winnipeg, Manitoba, remain relevant to anybody working with stream organisms: "We must, in fact, not divorce the stream from its valley in our thoughts at any time. If we do we lose touch with reality. The real lake is not a basin with two vertical sides as in the textbook. One that is like that, Loch Ness, is so out of line that it harbours monsters. Somewhere, in Australia, there must be a stream with a channel like a gutter, fed by runoff from a landscape paved like a parking lot. There, I predict, will be found the legendary river creature of the aboriginals – the Bunyip."

References

Aimer RD, Segedin BP (1985) Some aquatic hyphomycetes from New Zealand streams. N Z J Bot 23:273–299

Bandoni RJ (1981) Aquatic hyphomycetes from terrestrial litter. In: Wicklow DT, Carroll GC (eds) The Fungal Community. Marcel Dekker, New York, pp 693–708

Bandoni RJ, Koske RE (1974) Monolayers and microbial dispersal. Science 183:1079–1081

Bärlocher F (1981) Fungi on the food and in the faeces of *Gammarus pulex*. Trans Br Mycol Soc 76:14–19

Bärlocher F (1982) Conidium production from leaves and needles in four streams. Can J Bot 60:1487–1494

Bärlocher F, Murdoch JH (1989) Hyporheic biofilms – a potential food source for interstitial animals. Hydrobiologia 184:61–67

Bell DT, Sipp SK (1975) The litter stratum in the streamside forest ecosystem. Oikos 26:391–397

Bird GA, Kaushik NK (1981) Coarse particulate organic matter in streams. In: Lock MA, Williams DD (eds) Perspectives in running water ecology. Plenum, New York, pp 41–68

Blanchard DC, Parker BC (1977) The freshwater to air transfer of microorganisms and organic matter. In: Cairns JC (ed) Aquatic Microbial Communities. Garland, New York, pp 625–658

Botosaneanu L (1979) Quinze années de recherches sur la zonation des cours d'eau: 1963–1978. Bijdr Dierk 49:109–134

Byrne PJ, Jones EBG (1975) Effects of salinity on the reproduction of terrestrial and marine fungi. Trans Br Mycol Soc 65:185–200

Chamier A-C (1985) Cell-wall-degrading enzymes of aquatic hyphomycetes: a review. Bot J Linn Soc 91:67–81

Chauvet E (1989) Production, flux et décomposition des litières en milieu alluvial. Dynamique et rôle des hyphomycètes aquatiques dans le processus de décomposition. Thesis, Paul Sabatier University, Toulouse

Covich AP (1988) Geographical and historical comparisons of neotropical streams: biotic diversity and detrital processing in highly variable habitats. J N Am Benthol Soc 7:361–386

Cummins KW (1973) Trophic relations of aquatic insects. Annu Rev Entomol 18:183–206

Cummins KW (1974) Structure and function of stream ecosystems. BioScience 24:631–641

Cummins KW, Minshall GW, Sedell JR, Cushing CE, Petersen RC (1984) Stream ecosystem theory. Verh Int Ver Limnol 22:1818–1827

Dance KW (1981) Seasonal aspects of transport of organic and inorganic matter in streams. In: Lock MA, Williams DD (eds) Perspectives in running water ecology. Plenum, New York, pp 69–95

Deighton FC, Mulder JL (1977) *Mycocentrospora acerina* as a human pathogen. Trans Br Mycol Soc 69:326–327

Field JI, Webster J (1983) Anaerobic survival of aquatic fungi. Trans Br Mycol Soc 81:365–369

Fisher PJ, Petrini O (1989) Two aquatic hyphomycetes as endophytes in *Alnus glutinosa* roots. Mycol Res 92:367–368

Fisher J, Webster J, Petrini O (1991) Aquatic hyphomycetes and other fungi in living aquatic and terrestrial roots of *Alnus glutinosa*. Mycol Res 95:543–547

Freeze RA, Cherry JA (1979) Groundwater. Prentice-Hall, Englewood Cliffs, NJ

Gönczöl J (1975) Ecological observations on the aquatic hyphomycetes of Hungary, I Acta Bot Acad Sci Hung 21:243–264

Gönczöl J (1976) Ecological observations on the aquatic hyphomycetes of Hungary, II Acta Bot Acad Sci Hung 22:51–60

Gönczöl J (1989) Longitudinal distribution patterns of aquatic hyphomycetes in a mountain stream in Hungary. Experiments with leaf packs. Nova Hedwigia 48:391–404

Gönczöl J, Révay A (1983). Observations on the hyphomycetes inhabiting forest litter of Hungary. Acta Bot Acad Sci Hung 29:107–125

Goos RD (1978) Occurrence of *Triscelophorus monosporus* in upland stream sites on Oahu, Hawaii. Mycologia 70:188–189

Hedger J (1990) Fungi in the tropical forest canopy. Mycologist 4:200–202

Howe HF, Smallwood J (1982). Ecology of seed dispersal. Annu Rev Ecol Syst 13:201–228
Hynes HBN (1963) Imported organic matter and secondary productivity in streams. Proc XVI Int Congr Zool 4:324–329
Hynes HBN (1970) The ecology of running waters. University of Toronto Press, Toronto
Hynes HBN (1975) The stream and its valley. Verh Int Ver Limnol 19:1–15
Hynes HBN (1983) Groundwater and stream ecology. Hydrobiologia 100:93–99
Illies J (1961) Versuch einer allgemeinen biozönotischen Gliederung der Fliessgewässer. Int Rev Ges Hydrobiol 46:205–213
Illies J, Botosaneanu L (1963) Problèmes et méthodes de la classification et de la zonation écologique des eaux courantes, considérées surtout du point de vue faunistique. Mitt Soc Int Limnol 12:1–57
Ingold CT (1975) An illustrated guide to aquatic and waterborne hyphomycetes (Fungi Imperfecti) with notes on their biology. Freshwater Biol Assoc Publ No 30
Ingold CT (1976) The morphology and biology of freshwater fungi excluding phycomycetes. In: Jones EBG (ed) Recent Advances in Aquatic Mycology. Marcel Dekker, New York, pp 335–357
Iqbal SH, Webster J (1973) The trapping of aquatic hyphomycete spora by air bubbles. Trans Br Mycol Soc 60:37–48
Jones EBG, Byrne PJ (1976) Physiology of the higher marine fungi. In: Jones EBG (ed) Recent advances in aquatic mycology. Paul Elek, London, pp 135–175
Jones EBG, Oliver AC (1964) Occurrence of aquatic hyphomycetes on wood submerged in fresh and brackish water. Trans Br Mycol Soc 47:45–48
Kirk PW (1969) Aquatic hyphomycetes on wood in an estuary. Mycologia 61:177–181
Kohlmeyer J, Kohlmeyer E (1979) Marine Mycology. Academic Press, New York
Kuehn K, Koehn R (1988) A mycofloral survey of an artesian community within the Edwards Aquifer of central Texas. Mycologia 80:646–652
Likens GE (1984) Beyond the shoreline: a watershed-ecosystem approach. Verh Int Ver Limnol 22:1–22
Lock MA (1981) River epilithon – a light and organic energy transducer. In: Lock MA, Williams DD (eds) Perspectives in running water ecology. Plenum Press, New York, pp 3–40
Lodge DJ, Asbury CE (1988) Basidiomycetes reduce export of organic matter from forest slopes. Mycologia 80:888–890
Madsen BL (1972) Detritus on stones in small streams. Mem Ist Ital Idrobiol 29:385–403
Malmquist B, Nilsson LM, Svensson BS (1978) Dynamics of detritus in a small stream in southern Sweden and its influence on the distribution of the bottom animal communities. Oikos 31:3–16
Matthews WJ (1988) North American prairie streams as systems for ecological study. J N Am Benthol Soc 7:387–409
Mayack DT, Thorp JH, Cothran M (1989) Effects of burial and floodplain retention on stream processing of allochthonous litter. Oikos 54:378–388
Metzler GM, Smock LA (1990) Storage and dynamics of subsurface detritus in a sand-bottomed stream. Can J Fish Aquat Sci 47:588–594
Meyer JL (1990) A blackwater perspective on riverine ecosystems. BioScience 40:643–651
Meyer JL, McDowell WH, Bott TL, Elwood JW, Ishizaki C, Melack JM, Pecharsky BL, Peterson BJ, Rublee PA (1988) Elemental dynamics in streams. J N Am Benthol Soc 7:410–432
Michaelides J, Kendrick B (1978) An investigation of factors retarding colonization of conifer needles by amphibious hyphomycetes in streams. Mycologia 70:419–430
Minshall GW (1988) Stream ecosystem theory: a global perspective. J N Am Benthol Soc 7:263–288
Minshall GW, Petersen RC, Cummins KW, Bott TL, Sedell JR, Cushing CE, Vannote RL (1983) Interbiome comparison of stream ecosystem dynamics. Ecol Monogr 53:1–25

Minshall GW, Cummins KW, Petersen RC, Cushing CE, Bruns DA, Sedell JR, Vannote RL (1985) Developments in stream ecosystem theory. Can J Fish Aquat Sci 42:1045–1055

Moser JC, Perry TJ, Solheim H (1989). Ascospores hyperphoretic on mites associated with *Ips typographus*. Mycol Res 93:513–517

Mulholland PJ, Newbold JD, Elwood JW, Webster JR (1985) Phosphorus spiralling in a woodland stream: seasonal variations. Ecology 66:1012–1023

Müller E, Loeffler W (1982) Mykologie, 4th edn. Thieme Verlag, Stuttgart

Müller-Haeckel A (1982) Organic material leaving the river Ängerån and its fate in the estuary. In: Müller K (ed) Coastal research in the Gulf of Bothnia, Junk, The Hague, pp 81–104

Müller-Haeckel A, Marvanová L (1976) Konidienproduktion und -kolonisation von Süsswasser-Hyphomyzeten im Kaltisjokk (Lappland). Bot Not 129:405–409

Müller-Haeckel A, Marvanová L (1979a) Periodicity of aquatic hyphomycetes in the subarctic. Trans Br Mycol Soc 73:109–116

Müller-Haeckel A, Marvanová L (1979b) Freshwater hyphomycetes in brackish and sea water. Bot Mar 22:421–424

Müller-Haeckel A, Marvanová L (1982) Hyphomycetes in the mouth of the river Ängerån and their fate in the estuary. In: Müller K (ed) Coastal Research in the Gulf of Bothnia. Junk Publishers, The Hague, 105–114

Naiman RJ, Décamps H, Pastor J, Johnston CA (1988) The potential importance of boundaries to fluvial ecosystems. J N Am Benthol Soc 7:289–306

Newell SY (1976) Mangrove fungi: The succession in the mycoflora of red mangrove (*Rhizophora mangle* L.) seedlings. In: Jones EBG (ed) Recent advances in aquatic mycology. Paul Elek, London, pp 51–91

Nilsson S (1964) Freshwater hyphomycetes. Symb Bot Ups 18:1–130

Park D (1974) Aquatic hyphomycetes in non-aquatic habitats. Trans Br Mycol Soc 63:183–187

Park D (1982) *Varicosporium* as a competitive soil saprophyte. Trans Br Mycol Soc 78:33–41

Pennak RW (1971) Toward a classification of lotic habitats. Hydrobiologia 38:321–334

Pringle CM, Naiman RJ, Bretschko G, Karr JR, Oswood MW, Webster JR, Welcomme RL, Winterbourne MJ (1988) Patch dynamics in lotic systems: the stream as a mosaic. J N Am Benthol Soc 7:503–524

Prochazak K, Stewart BA, Davies BR (1991) Leaf litter retention and its implications for shredder distribution in two headwater streams. Arch Hydrobiol 120:315–325

Pugh GJF (1974) Fungi in intertidal regions. Veröff Inst Meeresforsch Bremerhaven Suppl 5:403–418

Ranzoni FV (1953) The aquatic hyphomycetes of California. Farlowia 4:353–398

Ranzoni FW (1979) Aquatic hyphomycetes from Hawaii. Mycologia 71:786–795

Reynolds CS, White ML, Clarke RT, Marker AF (1990) Suspension and settlement particles in flowing water: comparison of the effects of varying water depth and velocity in circulating channels. Freshwater Biol 24:23–34

Rossi L, Vitagliano-Tadini G (1978) Role of adult faeces in the nutrition of *Asellus aquaticus* (Isopoda). Oikos 30:109–113

Rounick JS, Winterbourn MJ (1983) The formation, structure and utilization of stone surface layers in two New Zealand streams. Freshwater Biol 13:57–72

Ryder Gl, Scott D (1988) The applicability of the River Continuum Concept to New Zealand streams. Verh Int Ver Limnol 23:1441–1445

Sanders PF, Webster J (1978) Survival of aquatic hyphomycetes in terrestrial situations. Trans Br Mycol Soc 71:231–237

Sanders PF, Webster J (1980) Sporulation responses of some 'aquatic' hyphomycetes in flowing water. Trans Br Mycol Soc 74:601–605

Sedell JR, Froggatt JL (1984) Importance of streamside forests to large rivers: the isolation of the Williamette River, Oregon, USA, from its floodplain by snagging and streamside forest removal. Verh Int Ver Limnol 22:1828–1843

Sedell JR, Naiman RJ, Cummins KW, Minshall GW, Vannote RL (1978) Transport of particulate organic material in streams as a function of physical processes. Verh Int Ver Limnol 20:1366–1375

Shearer CA (1972) Fungi of the Chesapeake Bay and its tributaries. III. The distribution of wood-inhabiting ascomycetes and fungi imperfecti of the Patuxent River. Am J Bot 59:961–969

Shearer CA, Crane JL (1971) Fungi of the Chesapeake Bay and its tributaries. I. Patuxent River. Mycologia 63:237–260

Shearer CA, Webster J (1985) Aquatic hyphomycete communities in the River Teign. I Longitudinal distribution patterns. Trans Br Mycol Soc 84:489–501

Singh N, Musa TM (1977) Terrestrial occurrence and the effect of temperature on growth, sporulation and spore germination, of some tropical aquatic hyphomycetes. Trans Br Mycol Soc 68:103–106

Speaker RW, Luchessa KJ, Franklin JF, Gregory SV (1988) The use of plastic strips to measure leaf retention by riparian vegetation in a coastal Oregon stream. Am Midl Nat 120:22–31

Sridhar KR, Kaveriappa KM (1987) Occurrence and survival of aquatic hyphomycetes under terrestrial conditions. Trans Br Mycol Soc 89:606–609

Sridhar KR, Kaveriappa KM (1988a) Occurrence and survival of aquatic hyphomycetes in brackish and sea water. Arch Hydrobiol 113:153–160

Sridhar KR, Kaveriappa KM (1988b) Survival of water-borne fungi imperfecti under non-aquatic conditions. Proc Indian Natl Sci Acad B54:295–297

Statzner B, Higler B (1985) Questions and comments on the River Continuum Concept. Can J Fish Aquat Sci 42:1038–1044

Statzner B, Higler B (1986) Stream hydraulics as a major determinant of benthic invertebrate zonation patterns. Freshwater Biol 16:127–139

Suberkropp K (1984) Effect of temperature on seasonal occurrence of aquatic hyphomycetes. Trans Br Mycol Soc 82:53–62

Suberkropp K (1991) Aquatic hyphomycete communities. In: Carroll GC, Wicklow DT (eds) The fungal community. Marcel Dekker, New York (in press)

Suberkropp K, Klug MJ (1980) The maceration of deciduous leaf litter by aquatic hyphomycetes. Can J Bot 58:1025–1031

Suberkropp K, Michelis A, Lorch H-J, Ottow JCG (1988) Effect of sewage treatment plant effluents on the distribution of aquatic hyphomycetes in the River Erms, Schwäbische Alb, F.R.G. Aquat Bot 32:141–153

Thakur SB (1977) Survival of aquatic hyphomycetes under dry conditions. Mycologia 69:843–845

Thomas K, Chilvers GA, Norris RH (1991a) Changes in concentration of aquatic hyphomycete spores in Lees Creek, ACT, Australia. Mycol Res 95:178–183

Thomas K, Chilvers GA, Norris RH (1991b) A dynamic model of fungal spora in a freshwater stream. Mycol Res 95:184–188

Thornton ML (1970) Transport of soil-dwelling aquatic Phycomycetes by earthworms. Trans Br Mycol Soc 55:391–397

Thornton ML (1971) Potential for long-range dispersal of aquatic Phycomycetes by internal transport in birds. Trans Br Mycol Soc 57:49–59

Townsend CR (1989) The patch dynamics concept of stream community ecology. J N Am Benthol Soc 8:36–50

Townsend CR, Hildrew AG (1984) Longitudinal pattern in detritivore communities of acid streams: a consideration of alternative hypotheses. Verh Int Ver Limnol 22:1953–1958

Trepl L (1987) Geschichte der Oekologie. Athenäum Taschenbücher, Frankfurt

Triska FJ (1984) Role of wood debris in modifying channel geomorphology and riparian areas of a large lowland river under pristine conditions: a historical case study. Verh Int Ver Limnol 22:1876–1892

Tudge C (1989) The rise and fall of *Homo sapiens sapiens*. Trans R Soc Lond B 325:479–488

Vannote RL, Minshall GW, Cummins KW, Sedell JR, Cushing CE (1980) The River Continuum Concept. Can J Fish Aquat Sci 37:130–137

Vogel S (1981) Life in moving fluids. Princeton University Press, Princeton

Waid JS (1954) Occurrence of aquatic hyphomycetes upon the root surfaces of beech grown in woodland soils. Trans Br Mycol Soc 37:420–421

Wall CJ, Lewis GB (1978) Survival of *Mycocentrospora acerina* conidia. Trans Br Mycol Soc 70:157–160

Wall CJ, Lewis BG (1980a) Infection of carrot plants by *Mycocentrospora acerina*. Trans Br Mycol Soc 74:587–593

Wall CJ, Lewis BG (1980b) Survival of chlamydospores and subsequent development of *Mycocentrospora acerina* in soil. Trans Br Mycol Soc 75:207–211

Wallace JB, Webster JR, Cuffney TF (1982) Stream detritus dynamics: regulation by invertebrate consumers. Oecologia 53:197–200

Ward GM, Cummins KW (1979) Effects of food quality on growth of a stream detritivore, *Paratendipes albimanus* (Meigen) (Diptera: Chironomidae). Ecology 60:57–64

Webster J (1975) Further studies on sporulation of aquatic hyphomycetes in relation to aeration. Trans Br Mycol Soc 64:19–127

Webster J (1977) Seasonal observations on aquatic hyphomycetes on oak leaves on the ground. Trans Br Mycol Soc 68:108–111

Webster J, Descals E (1979) The teleomorphs of water-borne Hyphomycetes from fresh water. In: Kendrick B (ed) The Whole Fungus, vol 2. National Museums of Natural Sciences, Ottawa, Canada, pp 419–451

Webster J, Descals E (1981) Morphology, distribution and ecology of conidial fungi in freshwater habitats. In: Cole GT, Kendrick B (eds) Biology of conidial fungi, vol 1. Acad Press, New York, pp 295–355

Webster J, Moran ST, Davey RA (1976) Growth and sporulation of *Tricladium chaetocladium* and *Lunulospora curvala* in relation to temperature. Trans Br Mycol Soc 67:491–495

Webster JR (1983) The role of benthic macroinvertebrates in detritus dynamics of streams: a computer simulation. Ecol Monogr 53:383–404

Webster JR, Benfield EF (1986) Vascular plant breakdown in freshwater ecosystems. Annu Rev Ecol Syst 17:567–594

Williams DD (1981). Migrations and distributions of stream benthos. In: Lock MA, Williams DD (eds) Perspectives in running water ecology. Plenum, New York, pp 155–207

Williams DD (1989) Towards a biological and chemical definition of the hyporheic zone in two Canadian rivers. Freshwater Biol 22:189–208

Wilson EO (1971) The insect societies. Harvard University Press, Cambridge

Winterbourn MJ (1986) Recent advances in our understanding of stream ecosystems. In: Polunin N (ed) Ecosystem theory and application. Wiley, New York, pp 240–268

Winterbourn MJ, Rounick JS, Cowie B (1981) Are New Zealand stream ecosystems really different? N Z J Mar Freshwater Res 15:321–328

Young SA, Kovalek WP, Del Signore KA (1978) Distances travelled by autumn-shed leaves introduced into a woodland stream. Am Midl Nat 100:217–222

3 Community Organization

Felix Bärlocher

3.1 Theories of Community Structure

3.1.1 Historical Background

In one of his stories, Baron von Münchhausen recounts how he was overtaken by darkness on a desolate, snowy plain. He tied his horse to what he assumed to be a tree stump, and fell asleep. In fact, the stump was the top of a church steeple, and as the snow melted during the night, von Münchhausen awoke next morning in the church yard and discovered his horse dangling from the top of the steeple. Von Baeyer (1986) compares the horse to classical physics: it is securely fastened to Newtonian mechanics. But what was once thought to be solid ground has melted away due to developments in quantum mechanics, relativity theory and deterministic chaos, leaving the discipline suspended over an abyss.

A theoretical population ecologist may be compared to a rider, who in the distance sees the top of a steeple piercing the snow drifts, and believes it to be a tree stump. He hopes to reach it, and attach his horse to it before nightfall. Maybe 20 years ago, many scientists were convinced that this stump, signifying classical physics, could indeed be reached shortly, and when reached, would provide a solid foundation for the understanding of biological communities. In the eyes of many ecologists, theoreticians or empiricists, these prospects have become increasingly remote.

They were based on the idea of a balanced, harmonic nature, which has developed independently in many cultures (Egerton 1973). Every biological species was believed to have its given, firm place in the scheme of things. These assumptions were not deeply disturbed by Darwin's writings; most biologists remained convinced that biological species and communities remain stable over long periods of time. After all, the majority of species had been in existence for thousands, perhaps millions of generations, yet their numbers had never increased to infinity, nor decreased to zero. Within the life span of a scientist, communities and species appeared to be in equilibrium. Drastic changes were assumed to be the result of human interference upsetting this balance. Biologists therefore occupied themselves primarily with charting and cataloguing the vast richness of nature. Other scientists contemptuously referred to this activity as "stamp collecting" as early as the beginning of this century, and Wordsworth describes a naturalist as "one that would peep and botanize upon his mother's grave" (McIntosh 1985). Labelled as scientific natural history, it led to the accumulation of relatively independent, non-integrated observations.

A new era was introduced in the 1920s by the physicist Lotka and the mathematician Volterra. They used mathematical equations to describe population growth of competitors, predators and prey species. Their aim was to put the postulated equilibrium, or balance of nature, on a firm theoretical basis. Their efforts drew a decidedly mixed reaction from biologists. Thus Elton, in a review of Lotka's work (quoted in McIntosh 1985), wrote: "Like most mathematicians he takes the hopeful biologist to the edge of a pond, tells him that a good bath will improve his work, and pushes him into the water and lets him drown."

Gause (1934), on the other hand, tested the predictions of Lotka and Volterra's equations by doing simple "bottle experiments" with microorganisms such as yeasts and *Paramecium*. The result which received the greatest attention was the fact that two species competing for the same resource in a homogeneous environment could not coexist for very long. This became known as Gause's competitive exclusion principle, and has been paraphrased as "Complete competitors cannot coexist" (Hardin 1960).

Central to Gause's principle is the assumption that birth and death rates are closely correlated with population size. With increasing density births decrease, and deaths increase. Not everybody accepted these relationships. Based on their experience with Australian invertebrates, Andrewartha and Birch (1954) concluded that population densities were generally small and had a variable effect on population growth. More recently, Strong (1986) labelled this phenomenon "density vagueness": within each population, there is an intermediate range of densities where population change is not distinctly related to density.

Even though the 1920–40s have recently been called the golden age of theoretical ecology (Scudo and Ziegler 1978), its influence on ecologists remained negligible (McIntosh 1985). This changed dramatically in the 1960s and 1970s, partly due to the increasing awareness of an impending environmental crisis. Suddenly, ecologists were at the centre of public attention and their advice and guidance were valued. This gave a welcome boost to their prestige and self-confidence. Most, however, were badly prepared. Politicians, the press and the public want clear, simple answers; any competent ecologist knows that simple capsule descriptions of nature are wrong, or at least misleading. As H.B.N. Hynes is fond of saying, ecologists cannot use a single number, ratio or index to describe nature any more than physicians can to describe the health of their patients.

Empirical, or field ecologists were further disconcerted by the attacks of more theoretically inclined colleagues. According to these, the lack of universal laws of ecology was due not so much to the complexity of nature, but rather to the immaturity of the science (and, by extension, of its practitioners). A determined attempt was made to apply the mathematical approach, so successful in physics, chemistry and molecular biology, to population ecology. Revitalization of theoretical ecology is closely linked to the contributions of G.E. Hutchinson and R.H. MacArthur, and their many students and associates. Their aim was to find simple rules of how species are packed into their environment, a process compared to crystal packing by MacArthur (1972). May (1974) envisioned the establishment of mathematical models that could serve as the "perfect crystals" of ecology.

An era of often overboarding optimism and enthusiasm followed, coupled with disdain for the descriptive, old-fashioned ways of field ecologists. According to Waterman (1965), only by developing theory "can biology outgrow the stigma of being a mere descriptive science and rise to the level of a rigorous intellectual discipline", and Pielou (1969) considered ecology to be essentially a mathematical subject. While in earlier years there had been a widespread perception that ecology suffered from too many undigested raw data, the new population theories led to a "constipating accumulation of untested models" (Schoener 1972), which led McIntosh (1985) to prescribe a purgative for the whole of ecology.

Elegance or originality of a model were, and often still are, more valued than its support by real ecosystems: "I am not going to discuss much data in this book. I am just going to discuss theory for the most part, and will leave it to each reader to fill in examples from his or her other studies in biology and ecology" (Yodzis 1989). To paraphrase Mark Twain: there is something fascinating about theoretical ecology. One gets such wholesale returns of conjecture out of such trifling investment of fact.

It was widely assumed that ecosystems were basically in an equilibrium, and that the number of coexisting species was largely determined by competition. This probably reflects Darwin's (1859) image of variously shaped wedges being pushed and hammered into a barrel as a metaphor for the way a set of species fit together to form a community. If one species changes just a bit, then another could, or maybe must, change as well, and all species would coevolve by a kind of domino effect.

Elton wrote (1933): "In a fairly limited area only a fraction of the forms that could theoretically do so actually form a community at any one time . . . The animal community really is an organized community in that it apparently has 'limited membership'." There have always been dissenting voices, both among zoologists (e.g. Andrewartha and Birch 1954) and among botanists. The most celebrated among the latter is undoubtedly Gleason (1926), who wrote: "The vegetation of an area is merely the resultant of two factors, the fluctuating and fortuitous immigration of plants and an equally fluctuating and variable environment. As a result, there is no . . . reason for adhering to our old ideas of the definiteness and distinctness of plant associations." Similarly, Elton (1930) was by no means convinced that natural communities resemble rigid mechanisms: "It is further suggested that if we knew enough about the ecological relations of the animals we could predict the effect of any interference, just as a clockmaker can work out the ultimate effect of the twirling of one wheel upon the rate of revolution of any of the others. At the same time it is assumed that an undisturbed natural community lives in a certain harmony, referred to as 'the balance of nature' . . . The picture has the advantage of being an intelligible and apparently logical result of natural selection in producing the best possible world for each species. It has the disadvantage of being untrue. 'The balance of nature' does not exist, and perhaps never has existed . . . The simile of a clockwork mechanism is only true if we imagine that a large proportion of the cog-wheels have their own main-springs, which do not unwind at a constant speed. There is also the difficulty that each wheel retains the right to arise and migrate and settle down in another clock, only to set up further trouble in its new home."

Among theoretical ecologists, however, the conviction prevailed that communities are tightly organized, that competition between species is the dominant organizing force and determines who is able to enter the "limited membership", and that it leads to recognizable and constant patterns in the occurrence of species.

On the basis of classical competition models, it was concluded that there could not be more consumer species than resource types, and consequently, that coexisting consumers cannot be too similar. Interspecific competition was considered a powerful evolutionary force that selects for adaptations resulting in species differing in their use of resources, thereby lessening the effects of competition. Geographic distribution of species was often believed to be determined by competitive pressures. This strong selection would therefore lead to evolutionary divergence. It has repeatedly been pointed out that there is an inherent problem of circularity in these statements. When coexistence is found, two presumably competing species are assumed to occupy different niches. When no competition is found, two species are said to have evolved apart due to competition in the past. Connell (1980) aptly named this invoking "the ghost of competition past".

The term niche has been used in ecology since the early 20th century. In a celebrated image, Elton (1927) suggested that when talking about a badger, an ecologist should include in his thoughts some definite ideas about the animal's place in the community, just as if he were talking about a vicar. Elton's concept of niche was somewhat ambiguous and included the role of a species, as well as an abstract function in a community which could be filled by any of a variety of species (the second usage resurfaced later in Cummins' definition (1974) of functional groups in stream invertebrates; (Chap. 2). Hutchinson's (1957) definition, by contrast, was based on a one species – one niche concept. He defined as fundamental niche the space enclosed by abiotic factors (e.g. temperature, humidity, pH values) that allow survival and reproduction of the members of a given species. As realized niche, he defined the space actually occupied by a species when interactions with other species (competitors, predators, pathogens, symbionts) are taken into account. Generally, the realized niche will be smaller than the fundamental niche (possible exception: mutualistic interactions).

During the heyday of theoretical ecology, its actual contributions to understanding nature were seldom critically examined. Many were intimidated by the brilliance of the mathematics or the arrogance of the theoreticians. There were important exceptions, however. Wilson (1969) warned of "an unholy alliance between population model builders and molecular biologists to exclude systematics and descriptive ecologists", and Hutchinson (1975) stated that "Modern biological education may let us down as ecologists if it does not insist . . . that a wide and quite deep understanding of organisms, past and present, is as basic a requirement as anything else in ecological education."

3.1.2 Recent Criticisms

3.1.2.1 Physics as a Model for Ecology

The criticism of ecological theory gathered steam in the late 1970s to early 1980s (e.g. Connell 1975, 1978, 1980; Wiens 1977; Simberloff 1980, 1982). It was facilitated by the fact that most of the earlier extravagant claims and promises had not been fulfilled. As prophets of ecological events, theoreticians were impressive primarily when predicting the presence from the past. An unfamiliar attitude of modesty appears in some of today's textbooks: "It is not inconceivable that the phenomena of ecology are just not amendable to such a rigorous approach. Theoretical ecology, while enjoying a measure of success has yet to fully prove itself" (Yodzis 1989).

What went wrong? On a fundamental level, it has been argued that classical physics is an unsuitable model for ecology. Simberloff (1980) classifies it as an idealistic science, where objects are viewed as imperfect embodiments of fundamental, unchanging structures. Differences between objects of the same type are interpreted as noise or disturbances. They can only confuse our attempts to see the ideal or true structure. Darwin and Wallace, of course, achieved their great breakthroughs by emphasizing individual variations, or noise. "What physicists view as noise is music to the ecologist; the individuality of populations and communities is their most striking, intrinsic, and inspiring characteristic, and the apparent indeterminacy of ecological systems does not make their study a less valid pursuit" (Simberloff 1980). As Cohen (1971) puts it "physics-envy is the curse of ecology", and Gould (1989) warns against any ranking of the "hard" versus the "soft" sciences, or the rigorously "experimental" versus the "merely descriptive": "Perhaps the saddest aspect of this linear ranking lies in the acceptance of inferiority by bottom dwellers, and their persistent attempt to ape inappropriate methods that may work higher up the ladder. When the order itself should be vigorously challenged, and plurality with equality asserted in pride, too many historical scientists act like the prison trusty who, ever mindful of his tenuous advantages, outdoes the warden himself in zeal for preserving the status quo of power and subordination. Thus, historical scientists often import an oversimplified caricature of 'hard' science, or simply bow to pronouncements of professions with higher status."

3.1.2.2 Does Competition Explain Community Structure?
Prevailing theory of the 1960s and 1970s also came under attack for its presumed overemphasis of interspecific competition for resources as the major structuring force of communities (Wiens 1977; Connell 1980; Simberloff 1982). It is probably true, but trivial, that most individuals of all species are occasionally prevented from acquiring a resource due to the presence of an individual of another species. This could be classified as competition. The crucial questions are whether such encounters are frequent enough, and their effects severe enough, eventually to exclude certain species from a community, and whether they result in recognizable patterns in community structures, such as resource partitioning (Weldon and Slauson 1986). The critics demanded, justifiably, that the primacy of competition in structuring natural communities would have to be demonstrated, not assumed. Patterns, which some have found in communities and attributed to competition, would have to be tested against the null hypothesis that they are due to random accumulation of species within migrating distance. These points are valid, but often extremely difficult or impossible to satisfy. Thus, an obvious field experiment would be to modify the abundance of one species and monitor abundances of its presumed competitors. For practical or ethical reasons, this is not always possible. Similarly, null hypotheses based on random reshuffling of existing communities may remain far from neutral when the ecological data are strongly conditioned by competitive interactions. This has been called the narcissus effect by Colwell and Winkler (1984): when sampling from a post-competition pool, the role of competition is likely to be underestimated, since its effect is already reflected in the pool. In computer generated island communities, in which they could adjust the severity of competition, they were frequently unable to reject the null hypothesis and conclude that competition had played a role, even though they knew better. Competition past may be as elusive as a ghost, but its effects can nevertheless be very real. Current competition may not be any easier to recognize: equations with n species can lead to virtually any pattern of population trajectories of the competitors, from single or multiple equilibrial, periodic, almost periodic or chaotic dynamics (May 1976; Roughgarden 1989). The outcome depends on how they compete, and Roughgarden (1989) concludes that a purely a priori exploration of competition theory is not likely to have much meaning.

Studies of competition, both empirical and theoretical, have often been restricted by convenient taxonomic boundaries, on the premise that competition is most intense among members of closely related species. In fact, assemblages such as lizards or seed-eating birds may represent only a small portion of the total community that is founded upon the use of some common resource pool. This restriction is evident in the definition of a guild (Root 1967): "a group of species that exploit the same class of environmental resource in a similar way." Darwin assumed that struggle is generally more severe between species of the same genus than between species of different genera. One of the more interesting developments in recent years has been the recognition that competition often occurs between taxonomically distant taxa, for example, between seed-eating rodents and ants (Brown and Davidson 1977), or, between aquatic hyphomycetes and stream invertebrates (Bärlocher 1981).

The last few decades, of often very heated discussions, have not led to any consensus of the frequency and severity of interspecific competition, but it seems fair to conclude that very few contemporary ecologists hold out much hope that competition theory will provide a generally valid explanation of species abundances and distributions (Strong et al. 1984; Roughgarden 1989).

3.1.2.3 Predation and Disturbance

Predators, or predator analogues, can decisively influence species richness of their prey organisms. Darwin (1859), with his lawn-mowing experiment, was the first to demonstrate that more species can coexist when their biomasses are kept low by harvesting. In 1966, Paine described what must rank as one of the most influential experiments in modern ecology. He studied communities inhabiting the marine rocky intertidal zone. In the absence of mortality, the system tended to culminate in a monoculture of a sessile mussel. This was prevented by a starfish, a "keystone predator". It continuously created empty space that could be colonized by weaker competitors. Its removal (or protection of the mussels by cages) led to a rapid decline in species diversity. Other agents of mortality included the settling of debris and ice scouring. Together, they can be subsumed as disturbance.

It is now generally accepted that predation, or, more generally, disturbance, can be a major organizing force in natural communities. The intermediate disturbance hypothesis by Connell (1978) states that species richness tends to be highest at intermediate levels of disturbance. It shouldn't come as a surprise that exceptions have been found. In some cases, the actual composition of the communities of rocky intertidals was determined primarily by what happened to settle out of the water column, and disturbance, or interactions of any kind, did not appear to exert any strong influence (Roughgarden 1989). The rule was also modified by theoreticians: Yodzis (1989) subdivided it into twelve different cases. An ecologist often experiences the same feeling one might get from observing the construction of a Cantor set or Cantor cheese, where a given proportion is removed from the centre of an original structure, the same proportion removed from the centres of the remaining subunits, and so on (Stewart 1989). One watches helplessly the decay of an apparently solid generalization, until nothing but holes and dust seem to remain.

A very instructive attempt to combine some of the more significant factors that influence community structure was published by Caswell (1978). He distinguished between local species diversity in cells harbouring sub-populations which are interconnected through migrations, and global diversity, the total of all species present in at least one cell. Predation often decreased local species richness and at the same time increased global species richness. Higher global species diversity of competitors was achieved by the maintenance of a non-equilibrium. Random, i.e. non-selective, harvesting of prey periodically emptied cells and allowed reinvasion by either of two competing prey species. It thus prevented monopolization by either of the two. Rather than asking for conditions at equilibrium, Caswell was interested in determining what happens before equilibrium is reached, and how long a non-equilibrium could persist. Computer simulation showed that such non-equilibria could be maintained for immense periods of time by periodic disturbances. Many of the properties of the model are now recognized as contributing to the community structure: the patchy distribution of organisms, environmental stochasticity, migrations, non-equilibria, competition and disturbances. It shows that even if equilibrium processes predominate on a small scale (and these might well be investigated satisfactorily in a bottle experiment), the sheer numbers of bottle equivalents in nature, combined with local fluctuations in migration and mortality, can largely invalidate any conclusions based on local observations. This seems to come close to what Salt (1979) defines as emergent properties: properties unique to a given level of biological organisation that are wholly unpredictable from observations of the isolated components of that level.

3.1.2.4 Deterministic Chaos

Theoretical ecologists are notoriously susceptible to the lure of new mathematical tools (toys?). Dayton (1980) claims that "Ecology often seems dominated by theoretical bandwagons driven by charismatic mathematicians, lost to the realization that good ecology rests on a foundation of natural history and progresses by use of proper scientific methods." The newest bandwagon appears to be deterministic chaos (May 1974, 1976, 1989; Stewart 1989; Sugihara and May 1990). This describes strictly deterministic processes leading to fluctuations that look like random noise. To distinguish between the two, one looks for the presence of strange attractors. Techniques developed for this purpose are extremely data-hungry. Thus, marine plankton samples, collected weekly over 19 years are considered a "relatively short run of real data" (Sugihara and May 1990). In this example, deterministic chaos accounted at best for about 50% of the variance, the rest was presumably due to additive noise. I don't anticipate a stampede of mycologists rushing to initiate similar long-term studies, with the prospect of reaching conclusions that only a mathematician could love. Nevertheless, the presence of chaotic dynamics may have important implications. If we follow concurrent population changes of several species, the amount or direction of changes may depend on their present locations. Sometimes all increase or decrease simultaneously, sometimes all may remain constant, or some may increase while others decrease (Roughgarden 1989). A naive observer will draw different conclusions about interactions among the observed species, even though they always followed the same rules. To date, chaos has been demonstrated most convincingly in simple systems (e.g. epidemics in human populations). May (1989) believes that it should be a fairly common phenomenon; field ecologists tend to be more skeptical (e.g. Myers et al. 1990). On a fundamental level, chaos theory is clearly inadequate, since it implies that populations change within a restricted space (the strange attractor). This does not seem to allow room for extinction or evolution.

3.2 Fungal Ecology

3.2.1 Recent Developments

Even to the casual observer it is obvious that theoretical ecology has had little influence on how mycologists study their organisms, and conversely, that investigations of fungi have had a negligible influence on the development of general ecological concepts and hypotheses (with the notable exception of Gause's work on coexistence and competition in yeasts). To correct this, Wicklow and Carroll (1981) urged mycologists to apply theoretical models to their organisms (ironically at the very time when theoretical ecology itself came under increasing attack from within). They felt that the low status of fungal ecology was "in part due to the reluctance of mycologists to step outside their own literature and attempt generalizations on the significance of their observations and experimental results." Cooke and Rayner (1984) wrote that fungal ecology has a wealth of descriptive information, but lacks a strong coherent framework. They attribute this to an inadequate understanding of fundamental fungal attributes, in particular, the behaviour of fungal mycelia. In other words, mycologists don't really understand fungi, and this lack of understanding has retarded the emergence of clear ecological principles.

Much of the more recent knowledge on the distribution and genetic structure of fungal populations in nature has been gained from observing wood-decomposing fungi, especially Basidiomycetes. This has clear methodological advantages: it is relatively easy to observe patterns of decay and discolouration and combine it with mapping and genetic analysis of naturally occurring mycelia. The persistence of the wood may allow time for a truly interacting community to develop and be maintained. Cooke and Rayner (1984) specifically mention that progress in fungal ecology has been impeded by the undue emphasis which has been placed on decomposition in soils. They make a valid point: some systems may simply be too intractable or confusing to yield any meaningful insights. It doesn't follow that results from the simpler system are valid, or useful, in the more complex system.

When two fungal mycelia meet, their hyphae can potentially fuse and thereby form cooperative units. Based largely on their experience with wood-inhabiting fungi, Cooke and Rayner (1984) concluded that this is not a general feature, even between conspecific mycelia, and less common than was assumed until recently. It is generally prevented by somatic incompatability. As a consequence, most individual mycelia of Ascomycetes are substantially homokaryotic. Basidiomycetes exist primarily as heterokaryons, derived from hyphal fusion between sexually compatible primary mycelia (homokaryons). Somatic incompatability between different secondary mycelia is common and manifests itself as a zone of sparse mycelium, cell lysis, the formation of pigmented zones, etc. Parasexuality (asexual exchange of DNA) appears to be the exception rather than the rule. The picture that emerges, then, is that the mycelia of a given species are generally subdivided into genetic subpopulations, which through somatic incompatability maintain their separate identities during vegetative growth, and compete independently for resources. These conclusions, of course, would be unremarkable in any other group of organisms; they nevertheless represent an important change of emphasis in mycology.

It is clear, however, that fungi continue to pose unique problems, and fungal ecologists often seem obsessed with discussions of methodology. At the very outset, there is the difficulty of identifying members of the community. Fungi may occur as mycelia, fructifications or spores. Only mycelia decompose substrates (though some hyphae may also be resting structures and both hyphae and spores can be dead). Fungal ecology is still dominated by a process-oriented point of view (viz. Christensen's, 1989, comment "Decomposition is everyman's response to what fungi do"), and it therefore seems desirable to identify those fungi that are present in an active mycelial state. Unfortunately, direct microscopic examination reveals mainly sterile hyphae whose identity or activity often cannot be directly established (the major exceptions are Basidiomycetes growing on wood, see above). Indirect methods must therefore be used. The simplest method, the dilution plate technique adopted from bacteriology, yields mostly colonies originating from spores, which do not participate in decomposition processes and may simply have been blown been from elsewhere (Garrett 1963). An alternative method was introduced by Warcup (1957). He isolated many fungi from soil by picking out individual hyphal fragments and plating them on sterile media. This approach undoubtedly yields more easily interpreted results, but it is extremely laborious and uncertain in most hands.

Mycologists may praise it, but hardly anybody ever uses it. Today, most rely on a combination of direct and indirect techniques, the latter usually based on plating out subsamples of the resource on suitable media.

Perhaps the best comment on direct microscopic observation and indirect isolation techniques remains that by Garrett (1952): "With the plate count one identifies what one cannot see (i.e. in situ), whereas with the direct method one sees what one cannot identify."

There are potential solutions, which have been used successfully in other habitats: immunological techniques to identify mycelium in situ, and uptake of radioactive substances to demonstrate metabolic activity (Wicklow and Carroll 1981; Mohamed et al. 1989). More recently, very specific and sensitive DNA analyses have been used to establish the presence or absence of different bacterial species or even strains (Giovannoni et al. 1990). An excellent review of problems associated with measuring fungal biomass and productivity, and potential solutions, was written by Newell (1991). These techniques hold some promise for fungal ecology. To date, none of these highly sophisticated and often expensive techniques have been applied to aquatic hyphomycetes, or indeed to an entire community in any habitat. But as pointed out by Kirby et al. (1990), the aim of ecological studies is not always the complete description of a community; sometimes an analysis of its dominant members, or the detection of certain patterns are sufficient. In these cirumstances, a comparative approach, based on simple techniques, is often adequate.

One of the early questions in fungal ecology was whether fungi participate in soil decomposition processes (Garrett 1963). It has since been established beyond any doubt that they do, but this early emphasis on growing and actively decomposing mycelia has continued to dominate fungal ecology at the expense of a more balanced interest in the entire life-cycle. Ecological work with fungal spores has become slightly suspect. This is clearly in contrast to current evolutionary thinking. It is generally agreed that an organism's success is best measured by the number of viable descendants it leaves during its life-time. Spore production is only the first step, but it can be assumed that those individuals of a species that produce more spores over a longer period of time will generally have a higher chance of establishing new colonies. This seems especially appropriate in the case of aquatic hyphomycetes; in soils, some Basidiomycetes may persist simply by growing through successive litter layers, which are replenished every fall and remain basically static (Cooke and Rayner 1984). Litter packs in streams are more ephemeral; there is a constant shifting and sorting out due to the current. I'm not aware of any evidence that individual leaves stick together due to fungal mycelium crossing from one leaf to the next; hyphae only seem to grow out of leaves in preparation for sporulation. The huge investment in conidium production made by aquatic hyphomycetes (Findlay and Arsuffi 1989) further increases the likelihood that dispersal from leaf to leaf occurs primarily through conidia. Acquisition of new territory by growth, rather than by spore settling, might be more pronounced on branches and twigs, where colonies usually persist for longer periods of time (Chap. 4). The organization of a Basidiomycete mycelium has been compared to that of a colony of army ants (Rayner and Franks 1987); with aquatic hyphomycetes, a comparison to aphids seems more appropriate.

It seems that a more mycocentric approach encompassing the "whole fungus" with all its stages (Kendrick 1979) is clearly overdue. For example, if a habitat contains only propagules of a fungal species that fails to establish itself, its impact on the ecosystem may be negligible. To the fungus, active elsewhere, these unsuccessful propagules may represent a significant drain on its reproductive potential. Conversely, if a fungus is actively growing in a habitat, but fails to reproduce, its effect on the ecosystem may be significant, but its continued existence in that place may well be in jeopardy.

3.2.2 Aquatic Hyphomycetes

3.2.2.1 Relative Importance in Leaf Decay

A sensible first step in any investigation dealing with fungi continues be an estimate of their impact. Since Ingold's (1942) study, it has been known that aquatic hyphomycetes grow and sporulate on leaves. By using bactericides and fungicides in laboratory experiments, and by direct microscopic examinations, it has been established that fungal biomass is generally much higher than bacterial biomass (Triska 1970; Kaushik and Hynes 1971; Findlay and Arsuffi 1989; van der Merwe and Jooste 1988; possible exception: beech leaves, Iversen 1973); metabolic activity and growth rates are also higher for fungi than for bacteria, though the differences are often much less pronounced.

In addition to aquatic hyphomycetes, Oomycetes and terrestrial fungi occur on leaves recovered from streams. Oomycetes are generally early colonizers, and decline rapidly (Newton 1971; Bärlocher and Kendrick 1974; Summerbell and Cannings 1981; Bärlocher 1990). Their coarse mycelium is easily distinguished from that of higher fungi. Among these, terrestrial fungi (genera such as *Aureobasidium, Cladosporium, Alternaria*) that colonize the phyllosphere of senescent leaves often persist. When leaves collected from streams were incubated under semi-natural conditions (ambient stream temperatures, nutrient poor media), aquatic hyphomycetes consistently outgrew and outreproduced terrestrial fungi (Bärlocher and Kendrick 1974; Suberkropp and Klug 1974, 1976; Chergui and Pattée 1988a). They were also more successful in decomposition experiments with sterile leaves at low temperatures. When leaves were sterilized before being introduced into a stream, terrestrial fungi could not be recovered in significant numbers until 4 to 5 months later; nevertheless, such leaves lost weight at the same rate as control leaves, and both groups had the same assemblage of aquatic hyphomycetes (Bärlocher and Kendrick 1974). These results suggest that terrestrial fungi, which are on the leaves before these fall into the stream, have little influence on the subsequent direction or rate of leaf decay, at least during the colder seasons and in clean, well-aerated streams. Possible exceptions might occur in areas where leaves become densely covered by sooty moulds prior to abscission. Even though the direct involvement of these fungi in the decay of leaves may be small, their biomass may have considerable influence on invertebrate feeding. Collier (1990) reported that up to 72% of the gut content of stonefly larvae in some New Zealand streams consisted of sooty mould mycelium.

Finally, aquatic hyphomycetes have been shown to produce a rich array of enzymes active towards the major leaf polysaccharides (Suberkropp and Klug 1980; Chamier 1985; Suberkropp 1991a, Chaps. 4, 6, 8). It seems clear that in temperate streams, they are generally the dominant fungal group involved in leaf decomposition. Still, the contribution of other fungi cannot be dismissed altogether. Godfrey (1983) reported weight losses of submerged alder leaf and filter paper disks caused by pure cultures of two terrestrial species, *Epicoccum nigrum* and *Cladosporium cladosporioides*. Species belonging to *Fusarium* and *Cylindrocarpon* (which would not be classified as aquatic hyphomycetes) can be common, sometimes dominant, throughout the year on leaves decaying in streams (Bärlocher and Kendrick 1974; Summerbell and Cannings 1981; Bärlocher 1982; Rossi et al. 1983; Chamier et al. 1984; Shearer and Webster 1985a). Finally, aero-aquatic fungi, which form conidia with special flotation devices and whose main habitat is stagnant water, are occasionally found on leaves in streams (Bärlocher and Kendrick 1974; Fisher and Webster 1981; Goos 1987).

Gessner and Schwoerbel (1989) recently reported that leaching of leaves (i.e. rapid removal of soluble substances upon immersion) is much reduced when fresh, rather than pre-dried leaves are used. At least in alder leaves, this affects the microbial community: colonization by aquatic hyphomycetes was delayed, while bacteria and Oomycetes appeared to be stimulated (Bärlocher 1991a). Most leaves enter the stream immediately after detachment from the tree (Fisher 1977), and may therefore be loaded with simple organic molecules during the early stages of decomposition. If the early results with fresh alder leaves can be confirmed and extended to other leaf species, it may well turn out that bacteria and Oomycetes play a larger role than now assumed.

3.2.2.2 Seasonal Occurrence

Having established that aquatic hyphomycetes play an important role in leaf decomposition, what do we know about their community structure? The most striking pattern is the pronounced seasonal fluctuations of most species. This was already noted in Ingold's (1942) first study: most species were more common from late summer to early winter than during the rest of the year.

Four methods have been used to estimate the abundance of aquatic hyphomycetes: (1) Counts of conidia suspended in the water. This can be done directly with an inverted microscope (Müller-Haeckel and Marvanová 1979), or, by using a membrane filtering technique (Iqbal and Webster 1973a). The conidia will be retained on the filter where they can be stained, counted and in many cases identified. This method gives no information on the origin of the propagules. Instead of being produced in the stream, they may have been washed in from the terrestrial surroundings, or dripped into the stream from the phylloplane of riparian trees (e.g. *Tripospermum* sp.; Tubaki et al. 1985). Up to 30 000 conidia per litre stream water have been counted (Webster and Descals 1981). (2) Persistent foam or scum may develop below rapids and trap numerous conidia (Ingold 1942). Such foam samples are usually biased towards morphologically more complex conidia, and sigmoid spores are underrepresented (Iqbal and Webster 1973b; Lindsey and Glover 1976). Gönczöl (1971) added detergents to stream water to increase foaming. (3) Artificial traps consisting of rosin-coated slides, cellophane, or plexiglass have been used by Bärlocher et al. (1977), Lindsay and Glover (1976) and Müller-Haeckel and Marvanová (1976). (4) Observation of fungi on substrata. This can be done with detritus collected at random from a stream bed, or, by exposing prepared, homogeneous samples for varying periods of time and at different seasons. In general, stream water and substrata samples indicate fairly abrupt changes in the abundances of individual species. In foam, conidia can be retained and may survive without germinating for several weeks to months. This tends to blur changes in actively growing or reproducing fungal populations.

There is general agreement that in temperate streams running through deciduous forests the fungal community reaches a clear peak in overall metabolic activity sometime in autumn. This is reflected, for example, by the rising number of conidia found in the stream water. Peaks beginning in October/November and lasting to December or even January were found in England (Iqbal and Webster 1973a, 1977; Sanders and Anderson 1979; Chamier and Dixon 1982; Shearer and Webster 1985b), Canada (Bärlocher et al. 1977; Michaelides and Kendrick 1978), Italy (Del Frate and Caretta 1983), France (Mercé 1987; Chauvet and Mercé 1988; Chauvet 1989). Switzerland and Germany (Bärlocher and Rosset 1981; Wood-Eggenschwiler and Bärlocher 1983) and in the USA (Metvalli and Shearer 1989). In Australia, concentrations of conidia were highest in late summer to early autumn (Thomas et al. 1989). In all these cases, the build-up of spores started a few weeks after leaf-fall, coinciding with a drop in temperature. The same pattern was found in the San Marcos River in Texas (Akridge and Koehn 1987), where the temperature varies little throughout the year (22 ± 1 °C). This makes sense, since it was shown that the production of conidia from submerged leaves reaches a maximum eight to twelve weeks after their introduction into streams (Bärlocher 1982). Such leaves, brought back to the laboratory, can release up to six conidia per μg weight in 48 h. On spruce needles, the increase in spore production is much slower and more gradual and does not reach a well-defined maximum. In addition, some conifer trees shed their needles more or less evenly throughout the year. As expected, the number of conidia in a Black Forest stream running through a pure spruce stand varied much less with season (Bärlocher and Rosset 1981).

In many studies cited above, a smaller peak was observed in spring, generally in the period between March and May. This seems to be correlated with the higher runoff in spring, though the causative mechanism is unknown. Presumably, considerable amounts of leaves will be washed in from the banks during periods of high rainfall or snow-melt. They may already be colonized by aquatic fungi or simply increase the supply of suitable substrates. Or, budscales, catkins, etc. might add to the fungal food base. An alternative explanation is that the increased waterflow will stimulate sporulation of the fungi present. This effect is well known from laboratory experiments (Webster 1975, see Chaps. 1, 7).

A similar pattern with two peaks was found in a subarctic stream (Müller-Haeckel and Marvanová 1979). However, conidium concentration in the stream started to rise before the autumnal leaf-fall, indicating that factors other than substratum quantity (maybe changing day length?) were involved.

The timing of conidium and species maxima in New Zealand streams varies between different regions, but generally seems well correlated with prevailing litter fall patterns (Aimer and Segedin 1985; Aimer 1989).

In a stream in Pakistan, the highest number of species was found in April (Iqbal and Bhatty 1979). The authors attribute this peak to heavy rainfall. Similarly, more hyphomycetes were encountered during the monsoon and post-monsoon periods than during dry periods in Indian streams, which in turn closely correlated with leaf deposition (Sridhar and Kaveriappa 1984, 1989; Chandrashekar et al. 1990; see Chap. 10).

While conidium concentrations in stream water are almost always higher in autumn than in spring, the opposite may be true if total numbers of conidia in transport are compared. In a Swedish stream, more than twice as many conidia were carried downstream in March compared to October (Müller-Haeckel and Rheinheimer 1983).

Studies in England (Lindsey and Glover 1976) and in Sweden (Müller-Haeckel and Marvanová 1976, 1979; Müller-Haeckel 1977) with artificial traps also revealed a pronounced peak in fall and a smaller one in spring. Observations on naturally occurring foam generally confirm this pattern (Nilsson 1964; Iqbal and Webster 1973b; Mer and Sati 1989), though heavy rains, throughout the year, can markedly increase the number of trapped conidia.

When natural substrata such as leaves, twigs, etc., are introduced into a stream and recovered periodically, a peak in fungal activity is observed after four to sixteen weeks, followed by a marked decline (Triska 1970; Willoughby and Archer 1973; Bärlocher and Kendrick 1974; Suberkropp and Klug 1976; Bärlocher 1982; Rossi et al. 1983). This could be due to substrate deterioration or to changes in the environment, such as increased water temperature which seems to stimulate activity of potentially competing terrestrial fungi (Bärlocher and Kendrick 1974). Thus, when autumn-shed leaves were introduced in summer instead of in fall/winter the conidium production from those leaves was reduced significantly (Bärlocher and Schweizer 1983). Similarly, Suberkropp (1984) found that in a Michigan stream fungi typical of tropical and subtropical areas dominated when leaves were introduced in the warmer months (see Sect. 3.2.2.4). Regardless of the causes for the fungal decline after a few weeks, the timing of leaf-fall in temperate regions will impose the familiar fluctuations in spore concentrations in the streams.

In tropical rainforests, leaf fall patterns are variable, and no useful generalizations are as yet possible. Leaf fall depends on the seasonal distribution of rain, episodic storm events and the types of forest species present (Walker 1985; Covich 1988).

Substrata randomly collected from streams in temperate regions are again most heavily colonized in the colder seasons (Ranzoni 1953; Conway 1970; Suberkropp and Klug 1974; Wood-Eggenschwiler and Bärlocher 1983). Seasonal variation on substrates in a Puerto Rico stream was less clear (Betancourt et al. 1987). On grass blades and on pine needles, but not on leaves, in streams in Pakistan, numbers and abundances of Ingoldian fungi were higher in the warmer season (Iqbal et al. 1979, 1980). In Indian streams, diverse fungal populations coincided with rainfall and leaf deposition patterns and not with water temperature (Sridhar and Kaveriappa 1984, 1989; Chandrashekar et al. 1990). Heavy rains were also followed by a sharp rise in species found on woody debris in a stream in Rhode Island, USA (Lamore and Goos 1978), and rainfall patterns had a greater influence than water temperature on fungal diversity in a South African stream (Sinclair et al. 1983).

Taken as a whole, these observations suggest that in most regions the aquatic hyphomycete community experiences one, maybe two distinct periods of increased vegetative growth and/or reproduction. The often drastic decline in between can make a species undetectable. Lower numbers of species in water and foam samples during the warmer seasons have been reported, among others, by Ferreira et al. (1981), Iqbal and Webster (1973a,b, 1977), and Müller-Haeckel and Marvanová (1979). On the other hand, Del Frate and Caretta (1983) found no clear seasonal fluctuations in species richness, and Iqbal and Bhatty (1979) detected a peak in April, after a heavy rainfall. Similarly, Willoughby and Archer (1973) concluded that species richness in their foam samples was largely determined by rainfall patterns.

On substrata introduced into streams, species richness is initially low, rises to a peak within a few weeks, then remains stable or declines slightly (Bärlocher and Kendrick 1974; Suberkropp and Klug 1976; Bärlocher 1980; Chamier and Dixon 1982; Chamier et al. 1984).

Despite these overall patterns of entire communities, there is considerable variation of how individual species vary through the seasons. This is further discussed under 3.2.2.4.

3.2.2.3 Competition

In a classic paper, Hairston et al. (1960) argued that all organisms taken together are limited by the amount of energy fixed, since the accumulation of fossil fuels is negligible, when compared with the rate of energy fixation through photosynthesis. In particular, the decomposers as a group must be food-limited since by definition they comprise the trophic level which degrades organic debris. It seems clear that the same reasoning can be extended to decomposers in streams, since, on an annual basis, there is no overall accumulation of organic detritus. Seasonal changes in aquatic hyphomycete spore numbers described above suggest that periods of overabundance of food are followed by a severe crunch lasting through the rest of the year (at least in temperate streams).

That resource limitation is probably the primary factor responsible for low fungal populations is also shown by the clear connection between the abundance or scarcity of riparian vegetation and fungal stream populations (e.g. Metvalli and Shearer 1989; Chergui 1990). There is nevertheless evidence that overall fungal productivity is also influenced by the water chemistry (Wood-Eggenschwiler and Bärlocher 1983; Suberkropp 1991b; Chap. 8).

Downstream transport by the current, consumption by invertebrates and breakdown by bacteria, Oomycetes, terrestrial fungi and aquatic hyphomycetes all contribute to the rapid disappearance of leaves in streams. There is little aquatic hyphomycetes could do to prevent removal by the current (short of anchoring the leaves to the sediment by mycelial cords; there is no evidence that this happens). It seems likely (but so far untested to my knowledge) that more retentive stream beds will allow higher fungal populations.

Interactions between aquatic hyphomycetes and invertebrates are complex (Chap. 6). Invertebrate activity sometimes substantially accelerates weight loss of leaves over controls where invertebrates are excluded; in other studies, no significant differences were found (Petersen and Cummins 1974; Prus 1981; Webster and Benfield 1986; Chergui and Pattée 1988b). The impact of invertebrate feeding appears to be minimal in large streams and rivers, in slow-flowing side-arms, as well as in streams where limited allochthonous input may be insufficient to support large populations of leaf-consuming detritivores (Webster and Benfield 1986; Chergui and Pattée 1988b). Within a stream, it can vary seasonally and between leaf species when invertebrates first consume the more nutritious leaves and then switch to less palatable types (Haeckel et al. 1973). In addition, invertebrates are highly selective; they discriminate between leaves or leaf patches colonized by different fungal species (Bärlocher and Kendrick 1973; Arsuffi and Suberkropp 1985; Chap. 6). The intimate intermingling of fungal species on a leaf (Shearer and Lane 1983; Chamier et al. 1984; Fig. 3.1) complicates matters: it could mean that invertebrates at least sometimes have to choose between combinations of species rather than between individual pure colonies. Nevertheless, selective feeding might be expected to influence the composition of the fungal community. In a series of studies (Bärlocher 1980, 1982; Bärlocher and Schweizer 1983), where animal consumption accelerated leaf decomposition, it changed the dominance patterns of early fungal colonizers (an expected effect of selective predators); their effect on late colonizers was more that of a competitor. They ate substrata before late arrivals had a chance to colonize them.

In any case, the few published studies indicate that invertebrate activity generally lowers fungal productivity (as measured by spore production; Bärlocher 1980, 1982; Bärlocher and Schweizer 1983; see Chap. 6). Potential counter-adaptations by fungi may involve "hiding" from invertebrates, for example by forming colonies that cannot be detected by the animal; deterring potential consumers by producing repellents; or the ability to pass unharmed through the digestive tract of detritivores (Bärlocher 1979, 1985). It has been known for many years that aquatic hyphomycetes vary in their palatability and that some are toxic to invertebrates (Chap. 6); in addition, conidia or conidium-producing structures of some fungi were shown to survive passage through the gut of *Gammarus pulex* (Bärlocher 1981). It remains unknown how common these phenomena are, and what impact they have on fungal survival. Again, the close intermingling of different species might complicate matters: the possibility exists that some fungi might gain (lose) from close proximity to an unpalatable (palatable) species. There is as yet no evidence of anything but random association between different fungal species (Chamier et al. 1984).

Little seems to be known about interactions between aquatic hyphomycetes and the other microbial groups on leaves (bacteria, Oomycetes and terrestrial fungi present on senescent leaves). Colonization by aquatic hyphomycetes was lower when leaves were introduced in summer rather than in fall/winter, which may be due to increased activity by terrestrial fungi (Bärlocher and Schweizer 1983; Puppi 1983). There is one report, based on association patterns, that indicates inhibition of bacteria by the fungi (Chamier et al. 1984).

It seems obvious, however, that competition between aquatic hyphomycetes does occur and can sometimes be severe. Whenever a spore lands on a piece of leaf already occupied by another fungus and partly degraded, the mycelium it produces will have access to fewer nutrients and its reproductive potential will be curtailed. This is shown, for example, by the decline of overall spore production from leaves after a few weeks of immersion in the stream, even when invertebrates were excluded (Bärlocher 1982, 1991a); it is further confirmed by the fact that preinoculated leaves

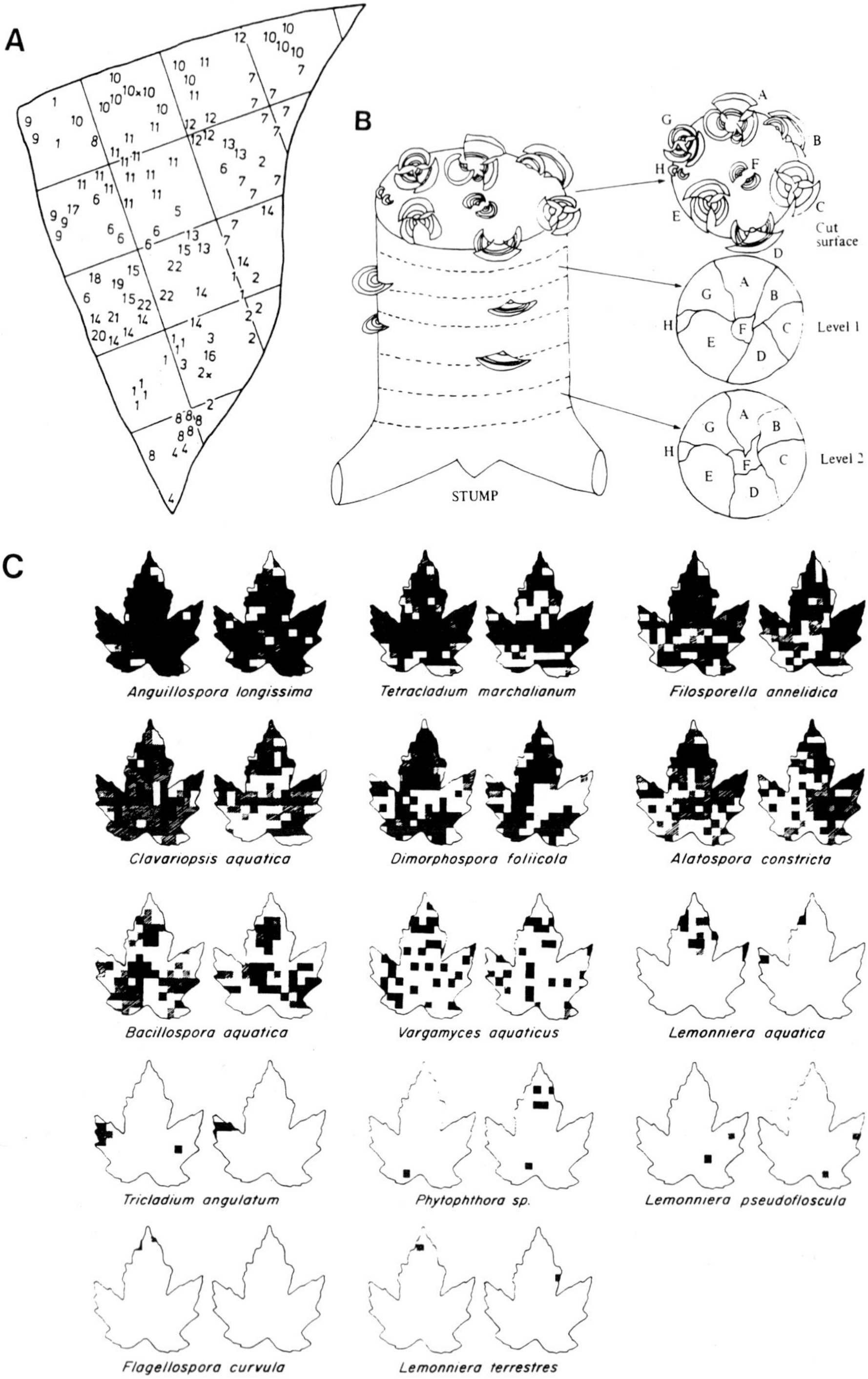
A
B
STUMP
Cut surface
Level 1
Level 2
C
Anguillospora longissima
Tetracladium marchalianum
Filosporella annelidica
Clavariopsis aquatica
Dimorphospora foliicola
Alatospora constricta
Bacillospora aquatica
Vargamyces aquaticus
Lemonniera aquatica
Tricladium angulatum
Phytophthora sp.
Lemonniera pseudofloscula
Flagellospora curvula
Lemonniera terrestres

were colonized to a lesser degree than fungus-free leaves (Suberkropp 1984; Rosset and Bärlocher 1985). Fungal activity thus contributes to the yearly crunch experienced by aquatic hyphomycetes: those leaves that do not disappear entirely are made less suitable for growth and reproduction.

From these observations, it appears that early resource capture is of paramount importance for an aquatic hyphomycete. Once established on a leaf, two basic strategies seem possible: a defence of the acquired resource (labelled "combat" by Cooke and Rayner 1984), or, channelling much of the newly acquired energy into rapid production of propagules in order to capture yet more leaves. The available evidence strongly suggests that rapid reproduction dominates. Up to one million conidia can be produced from a single leaf in 48 h (corresponding to 5–8 conidia per μg leaf substance, Bärlocher 1982, 1991a). Carbon released as spores can range from 18–122% of hyphal carbon remaining on the leaf (Findlay and Arsuffi 1989), and can account for 2–4% of the total carbon loss from leaves during the first 15–30 days of degradation. A similar estimate was published by Suberkropp (1991b), who showed that sporulation closely follows growth.

By contrast, defence of acquired resources appears relatively unimportant. As already pointed out by Ingold (1942), a leaf collected from the stream is typically colonized by 5–12 different species. This diversity seems to persist through much of the leaf's decay. Greathead (1961) found as many as seven species growing on one cm^2 of a skeletonized *Rubus* leaf. Closer examination of areas as small as 4 mm^2 showed extensive overlapping of colonies belonging to different species (Chamier et al. 1984; Shearer and Lane 1983). This is in striking contrast to the maintenance of large territories observed in many terrestrial Basidiomycetes (Cooke and Rayner 1984; see Fig. 3.1).

Fig. 3.1A–C. Different scales are necessary to characterize the community structure of terrestrial Basidiomycetes and aquatic hyphomycetes. **A** Mycelial isolates from an oak stand of the cord-forming agaric *Tricholomopsis platyphylla*. Based on mycelial interactions, 22 mutually incompatible types were found. Isolates belonging to the same mycelial type are generally found in close proximity. Size of quadrats = 160 m^2 (Thompson and Rayner 1982). **B** Simplified diagram of a tree stump colonized by *Coriolus versicolor*. The stump is cut into transverse slices. The position of the different decay columns is delimited by narrow dark zones, and can be correlated with that of fruit bodies at the surface. *Letters* indicate different mycelia. Isolates from separate decay columns were invariably antagonistic in culture, those of the same column merged readily. Diameter of stump approx. 10 cm (Rayner and Todd 1982). **C** Areas within a single silver maple leaf in which different species of aquatic hyphomycetes occurred as developing (*black squares*) or loose (*gray squares*) conidia. *Left* member of each leaf pair represents top surface, *right* member represents bottom surface. Several species may occur within the same quadrats of 36 mm^2. Nothing is known concerning the occurrence of genetically distinct individuals within this community. Microscopic examination does not reveal any zones of interactions (Shearer and Lane 1983)

Some aquatic hyphomycete species were found to occur in almost all quadrats of a leaf ($36\,mm^2$, Shearer and Lane 1983). It is not known whether they formed a single colony or consisted of separate small colonies of genetically distinct individuals. Recent experiments with conspecific strains isolated from widely separated areas suggest that somatic incompatability is much less pronounced than in wood-decomposing Basidiomycetes and Ascomycetes (Bärlocher 1991b). Of course, it would be premature to conclude that this results in frequent asexual exchange of genetic material.

In a study of interactions between eight different species, Khan (1987) found no evidence of inhibition at a distance, nor of hyphal interference. Antagonistic effects apparently expressed themselves by rapid colonization of available substrate and hence deprivation of the slower growing fungus. Mutual aggression among aquatic hyphomycetes thus appears to play a minor role. It cannot be discounted altogether, however. In vitro, close to 100% of all interactions between 25 species resulted in lowered growth of one or both members of an interacting pair (Shearer and Zare-Maivan 1988). Eight species produced zones of inhibition suggesting that they release diffusible antifungal substances. Aggression may play a more important role when the fungi grow on wood (Chap. 4), indicating considerable flexibility in the use of different strategies.

There is also the possibility that competition becomes effective before mycelial interactions can occur. Nilsson (1964) suggested that aquatic hyphomycetes are relatively rare on elm leaves because of their "hairy" surface. Similarly, when a leaf is covered by a dense forest of conidiophores, or, during the first few days of submersion, by a dense tangle of Oomycete mycelium and sporangia, a spore may be trapped before it reaches the leaf surface. If it germinates it may not have enough endogenous nutrients to reach the leaf blade and its nutrients.

Overall, aquatic hyphomycetes could be characterized as having adopted a stress-tolerant, ruderal strategy: the ability to grow at low temperatures, to digest relatively complex substrates, a short life span, coupled with high reproductive potential and little evidence for open aggression (Pugh 1980; Cooke and Rayner 1984). It should not be forgotten, however, that leaves represent only one phase in the life of these fungi. When some of the same species grow on wood, they often display a combative behaviour more suitable for long-lived resources (Chap. 4). The significance of these various types of substrata for the continued survival of a species in a stream is unknown: would it become extinct if all branches and twigs were removed from a given area? Would it increase if wood were added? Presumably there is some feedback in the sense that a successful fall/winter season on leaves allows recolonization of sufficient woody substrata to ensure sufficient inocula for the next leaf fall. The involvement of sexual reproduction in this annual cycle remains unknown. In any case, events occurring outside leaves, or even outside streams (Chap. 2) may influence coexistence of aquatic hyphomycetes.

3.2.2.4 Niche Partitioning

Differences in Spore Shape and Size. It is often believed that severe competition should accentuate differences between competitors. The most striking characteristic of aquatic hyphomycetes is their spore shape, an obvious example of evolutionary convergence, rather than divergence. Spore volume, however, varied by a factor of more than 50 among 27 species found on oak leaves (Bärlocher and Schweizer 1983). There may be some trade-off between the potentially increased production of cheaper, smaller spores, and their lower reserves of nutrients to tide them over the critical germination period. Spores apparently also differ in their affinity to surface structure or chemistry (Bärlocher et al. 1977): when glass-slides were coated with rosin, settling and germination of spores were higher than on control slides. The increase varied with species, suggesting that aquatic hyphomycetes possess an automatic substrate-selection mechanism which might help to counteract the apparently total passivity of their conidium dispersal mechanism. If this plays any role at all, it is not apparent at the level of entire leaves. Initial distribution of colonies on alder leaves was random (Chamier et al. 1984).

Temperature Preferences. In temperate regions, *Triscelophorus monosporus*, *Flagellospora penicilloides* and *Lunulospora curvula*, are restricted to warmer seasons (Ingold 1942; Conway 1970; Gönczol 1971; Suberkropp and Klug 1974; Suberkropp 1984) and there is little doubt that their primary distribution is in tropical/subtropical countries (Iqbal et al. 1979; Ranzoni 1979; Chap. 10). Temperature responses of growth and sporulation are broadly linked to geographical distribution (Nilsson 1964; Bärlocher and Kendrick 1974; Koske and Duncan 1974; Singh and Musa 1977; Suberkropp and Klug 1981; Suberkropp 1984). In *L. curvula* and *F. penicilloides*, growth already stops at 5 °C (Suberkropp 1984; Webster et al. 1976; no information is available for *T. monosporus*). But the relative success of winter or summer species in a Michigan stream could not be explained solely by temperature growth responses (Suberkropp 1984). Presumably, biological interactions modified temperature responses.

There are many other species restricted to tropical and subtropical areas, but no detailed information on their seasonal behaviour is as yet available. On the other hand, some species appear to be less common outside temperate regions (e.g. *Alatospora acuminata, Clavariopsis aquatica, Flagellospora curvula*; Nilsson 1964; Müller-Haeckel and Marvanová 1979).

Substratum Selectivity. An obvious mechanism to achieve niche separation would be to use the leaf's resources differently. Pushed to its extreme, substrate specificity could lead to different fungal species occurring on different leaf species. There is only one report where this is strictly true: in Mosedale Beck, where grass blades are the sole natural substrata, *Tricladium giganteum* was the only aquatic hyphomycete (Willoughby and Minshall

1975). Generally, there is little specificity if only absence/occurrence on different leaves, or even conifer needles, is determined (Ingold 1975; Webster and Descals 1981; Aimer 1989). Some leaves quickly accumulate a diverse fungal flora (e.g. alder in temperate regions, Ingold 1975; or coffee, rubber and banyan in tropical regions, Sridhar and Kaveriappa 1988; in Himalayan streams, seeds of *Tagetus erecta* Linn. were found to be the best bait, Mer and Khulbe 1981). In others, colonization proceeds slowly but the total number of species during the residence time may equal that of more quickly decaying leaves (e.g. oak leaves and conifer needles, Bärlocher 1982).

It is well established, however, that changes in the riparian flora often coincide with changes in the aquatic hyphomycete community (e.g. Gönczöl 1975, 1987, 1989; Bärlocher 1982; Wood-Eggenschwiler and Bärlocher 1983; Chauvet 1989; Thomas et al. 1989), though it is often difficult to disentangle the effects of substrata and water chemistry. Nevertheless, when leaves of different species are collected from the same stream section dominance patterns in the fungal communities of the leaves usually differ (Gönczöl 1975, 1989; Suberkropp and Klug 1976; Chamier and Dixon 1982; Bengtsson 1983; Rossi et al. 1983; Shearer and Lane 1983; Sridhar and Kaveriappa 1988, 1989).

Chauvet (1989) attributes scarcity of fungal colonization in large rivers to the scarcity of conidia. In the words of Roughgarden (1989), "... a community reflects both its applicant pool and its admission policies". Several authors observed that there appears to be little connection between types and numbers of spores carried in stream water, and the species which eventually colonize introduced leaves, twigs or wood blocks, or those found on randomly samples leaves (e.g. Chamier and Dixon 1982; Iqbal and Webster 1977; Sanders and Anderson 1979; Shearer and Lane 1983; Willoughby and Archer 1973). This might indicate a very selective admission policy. The phenomenon is partly due to presence of spores swept in from the outside (with morning dew, or from the terrestrial surroundings, Chap. 2); partly due to the misleading comparison of propagules with propagule-producing structures. A more meaningful comparison would be between the stream spora present during the decomposition of a leaf, and the spora released from that leaf during its decomposition. When this was done with leaves and conifer needles in four streams, there was a clear correspondence between the riparian vegetation, stream spora and spora released from leaves and needles (Bärlocher 1982). For example, oak leaves reproduced the stream spora of a stream running through a mixed deciduous forest with a similarity of up to 70%, and that of a stream running through a spruce forest with a similarity of only 33%.

Thus, most aquatic hyphomycetes colonize a wide range of substrata. Variations in their frequencies of occurrence on different leaf species nevertheless suggest that the substratum's composition selectively inhibits or stimulates colonization by different fungal species. This could already occur during spore settling and germination, or, it could be based on the range of compounds present in the leaf. These include nutrients (e.g. cellulose, hemicelluloses, lignins) and growth inhibiting substances (e.g. phenolics). Different species vary in their susceptibility to soluble substances present in freshly shed leaves (Bärlocher and Oertli 1978; Bärlocher 1990) or wood (Gunasekera and Webster 1983). Similarly, latex of rubber, calotropis, banyan, jack and papaya leaves had variable effects on growth and sporulation of five aquatic fungi (Sridhar and Kaveriappa 1987). *Heliscus lugdunensis* (an early colonizer of wood) has been reported to be less susceptible than most other species (Gunasekera and Webster 1983). I'm not aware of any other example linking tolerance of leaf extracts with success in colonization.

Differential Use of Polysaccharides. The vast majority of aquatic hyphomycetes use pectic polysaccharides and hemicelluloses (Chamier 1985; Chaps. 4, 6, 8). Differences are most pronounced when degradation of cellulose (particularly native or crystalline cellulose) is compared. Chamier et al. (1984) found strong cellulolytic activity in five of the seven dominant species. A sixth dominant had a strong activity towards cellobiose, a disaccharide released during cellulose breakdown, and thus may profit from excess cellulolysis by the other fungi. This suggests that there is some degree of niche partitioning. This is further supported by the observation that simultaneous inoculation of sterile leaves with several fungal species increases the rate of weight loss (Bärlocher and Kendrick 1974), suggesting that different species use different leaf compounds. At present, however, enzymatic differences do not provide a firm basis which might explain presence or absence, let alone dominance, of certain species. While *Tetracladium marchalianum*, *Heliscus lugdunensis* and *Lemonniera aquatica* consistently show strong cellulolytic activity (Suberkropp and Klug 1981; Suberkropp et al. 1983; Chamier et al. 1984; Zare-Maivan and Shearer 1988; Abdullah and Taj-Aldeen 1989), others, such as *Tricladium splendens*, *Lunulospora curvula* and *Tetrachaetum elegans* gave variable results (Suberkropp et al. 1983; Chamier et al. 1984; Chandrashekar and Kaveriappa 1988; Abdullah and Taj-Aldeen 1989). This may be due to differences in techniques, or, it could be that enzymatic activities are not constant within a given species, but vary from stream to stream. If the latter is the case, it suggests that a given species does not fill a unchanging niche defined by its enzymatic equipment. It could be that a given niche is filled by different species in different streams, or, that niche partitioning by enzymatic equipment is relatively unimportant in structuring the community. In any case, there is no indication that measuring enzymatic activities in vitro, or weight losses caused by pure cultures consistently allow predictions about the early success of a fungus on a given leaf.

Interactions Between Season and Substratum Quality. To complicate the issue further, the leaf's composition changes during decay. Generally, cellulose and hemicelluloses decline and lignin increases (Suberkropp et al. 1976; Rosset et al. 1982). If niche partitioning due to selective use of certain leaf components is important, one would therefore expect a succession of species becoming dominant at different stages of decomposition. Since decomposition may extend over several months and therefore seasons, the effects of substratum and season have to be distinguished. In a study of a spring-fed stream with a constant temperature of 22 °C (Akridge and Koehn 1987), no clear substratum preferences (eight leaf species) were found. There was no clear pattern of "early" species being replaced by "late" species (at least as far as spores in water and foam are concerned). All fungi reacted in a similar, predictable way to the autumnal leaf fall and the gradual disappearance of leaves. *Lunulospora curvula* and *Tetracladium marchalianum* remained dominant throughout the year. The community was unusually species poor; over the entire year, only 13 species were found. This may have been related to the constant temperature (only species sporulating at 22 °C would have been likely to survive), and the alkaline water (pH 7.2–9.4).

A more extensive study was published by Suberkropp (1984). When new leaves were introduced throughout the year at four week intervals, a "winter assemblage" (*Flagellospora curvula*, *Lemonniera aquatica* and *Alatospora acuminata*) dominated in the colder months. It was replaced by a "summer assemblage" (*Flagellospora penicillioides*, *Lunulospora curvula*, *Triscelophorus monosporus*) in spring and summer. Other species (e.g. *Anguillospora* sp., *Tetracladium marchalianum*) showed little seasonal variation. These patterns were clearly related to temperature preferences: the summer species had higher minimum and optimum temperatures. But the connection was not straightforward. In the coldest stream section, the summer species were absent even when the temperature rose above the minimum values. When they were transplanted to this site on leaves, they maintained dominance during the warmer season. On the other hand, *Tetracladium marchalianum* showed a temperature growth curve similar to the winter species but continued as a co-dominant on hickory leaves throughout the summer. Suberkropp (1984) suggested that interspecific interactions modify the temperature response. The dominance pattern remained essentially the same during three successive summers. Augusta Creek is a hardwater stream; in two studies, a total of 14 aquatic hyphomycetes have been identified (Suberkropp and Klug 1976; Suberkropp 1984).

Abel (1985) introduced oak leaves (*Quercus petraea*) at two month intervals, and determined spore production from leaves collected after two months incubation. In a hardwater stream (temperature between 5 and 14.5 °C), *Alatospora acuminata* remained dominant and *Clavariopsis aquatica* was a near second during 24 successive months (they exchanged ranks only once in 13 samples). *Tricladium angulatum*, *Tetracladium marchalianum* and *Heliscus lugdunensis* also were consistently among the top five. Compared to a study conducted two years earlier (Bärlocher 1982),

Clavariopsis aquatica and *Tetracladium marchalianum* switched places, otherwise there were no changes in dominance patterns. In the same stream, similar dominance patterns were found on larch and spruce, as well as in bags of fine mesh (excluded macroinvertebrate feeding) or coarse mesh (allowed feeding).

In a softwater stream (temperature between 0.9 and 11.8 °C), patterns were more diverse and more variable (Bärlocher 1982; Abel 1985). *Flagellospora curvula* was dominant during September/October of two successive years, but it also remained among the top four species through the summer months. Three other species (*Clavariopsis aquatica*, *Pyricularia aquatica*, *Articulospora tetracladia*) were the dominant spore producers on at least one occasion. There were no obvious winter and summer assemblages. This may have been due to the low water temperature (≤ 11.8 °C). Not surprisingly, *Lunulospora curvula*, *Triscelophorus monosporus* and *Flagellospora penicillioides* were never observed. On the other hand, *Alatospora acuminata* and *Lemonniera aquatica*, generally considered to be typical cold-water species, were never particularly common. Compared to Bärlocher's study (1982) in the same stream two years earlier, there were pronounced differences. On oak, *Clavatospora longibrachiata*, *Clavariopsis aquatica*, *Articulospora tetracladia*, *Alatospora acuminata*, and *Dactylella aquatica* had been the dominant spore producers. Substratum type (larch and spruce) and mesh size strongly influenced dominance patterns.

In contrast to Suberkropp's study (1984), Abel (1985) found weak or no evidence for seasonal succession; dominance patterns during two successive years were more stable in the hardwater stream than in the softwater stream (total number of aquatic hyphomycete species were 37 in the softwater and 18 in the hardwater stream, Bärlocher and Rosset 1981).

There are four other cases where the community structure of the same stream was examined in more than one year. In the River Bourne (softwater, total number of aquatic hyphomycete species = 26), only three out of six dominant species were identical in two studies conducted five years apart (Chamier and Dixon 1982; Chamier et al. 1984).

In the River Teign, Shearer and Webster (1985a–c) the following five species were dominant on alder leaves introduced on 28 November, 1979: *Geniculospora inflata*, *Clavatospora longibrachiata*, *Alatospora constricta*. *Lemonniera terrestris* and *Tetrachaetum elegans*. Ten year later, on alder leaves introduced on 7 November, only *C. longibrachiata* and *T. elegans* were still among the top seven species. *Anguillospora filiformis*, which was not found in the earlier study, was now the dominant species (Bärlocher 1991a).

In the Morgó stream (Hungary), there were considerable changes in rankings between fall/winter 86/87 and 87/88. Generally, only 2–3 of the top five species remained identical (Gönczöl 1989; total number of species not given).

In the Western Ghat forest (India), the top three species remained identical in two successive years in a stream with 36 species; in a neighbouring stream with 20 species, the top five species remained the same (Chandrashekar et al. 1990).

With the exception of the studies by Abel (1985) and Suberkropp (1984), the primary aim was not to compare fungal communities during successive years. Inevitably, there were some differences in techniques and timing, and any conclusions must be preliminary. It does appear that the stability of dominance patterns of aquatic hyphomycete communities varies between streams. Possibly, it decreases in more diverse communities. Intuitively, this makes sense: in an environment with more potential competitors it should be more difficult to stay consistently on top.

In a study dealing with the same topic, Townsend et al. (1987) compared invertebrate communities at 27 stream sites in southern England in 1976 and 1983. They found very substantial changes, both in overall species composition and species ranking. The most persistent assemblages were found at upstream sites with low discharge range or low, stable pH. They speculate that "in the less variable and perturbed streams, communities are rather more predictable and structured more prominently by deterministic, biotic interactions." At this stage, it would be premature to conclude that stable fungal communities (which may be species-poor) are structured by biotic interactions, whereas species-richer, less stable communities are not. But it seems worthwhile to initiate long term studies of fungal communities in various streams with a standardized technique.

In many other studies, leaves were collected randomly throughout the year, or introduced during or shortly after the main leaf-shedding season and recovered at various intervals. Many different leaves or needles were recovered from different stream types, and different techniques were used to establish dominance patterns. Not surprisingly, there is considerable variation. In a total of 20 studies, all conducted in temperate regions, 29 different fungal species reached one of the top five ranks at least once (Table 3.1; the top five ranks were based on an estimate of their cumulative presence during the entire experimental period). There is considerable variability in the timing of their appearance. Four species behaved fairly consistently in most studies. The best example is *Lunulospora curvula*. If present at all, it is always more common in the warmer season. A similar decline in later stages of decay is shown by *Flagellospora curvula* (one exception). *Alatospora acuminata*, on the other hand, is almost always more common in later samples (coinciding with colder stream temperature). *Tetracladium marchalianum* and *Clavariopsis aquatica* are generally common throughout. The other species do not exhibit a clear preference for certain stages of decay. It is possible to dismiss this absence of patterns as an artifact, due to different techniques, substrata, and streams. The fact remains that, at this point, there is little indication that most aquatic hyphomycete species are tied to specific stages in the breakdown of leaves. The possibility remains that during leaf decay well-defined, different niches open sequentially and predictably, and happen to be filled by different species under different conditions. This would lead to succession.

Table 3.1. Dominant fungal species and their temporal patterns during leaf decay in temperate streams

Alatospora acuminata	Dominant throughout study [1, 3, 5, 6, 10][a]
	Absent in early samples, later stable [9, 18]
	Tends to increase in later samples [3, 7, 16, 17, 19]
	Most common in late samples [2]
	No information [11, 15]
Anguillospora filiformis	Very dominant in early samples, later declines [4]
	Increases in mid-decay [2]
Anguillospora longissima	Rapidly declines after early dominance [7]
	Most common in mid-decay [18]
	Increased in later samples [2, 6, 12, 20]
	Dominant throughout [5, 10, 14]
	No information [11]
Anguillospora pseudolongissima	Late samples [13]
	Present throughout [13]
Articulospora moniliformis	Present throughout [20]
Articulospora tetracladia	Most common in early and late samples [8]
	Most common in mid-decay [2]
	Increases in later samples [4, 16]
Campylospora chaetocladia	Sporadic [20]
Clavariopsis aquatica	Common throughout [1, 9, 10, 14, 17, 19]
	More common in later samples [2, 16]
	No information [15]
Clavatospora longibrachiata	Early [13]
	Most acommon in mid-decay [1]
	More common in later samples [16]
	Only in late samples [10]
Filosporella annelidica	No information [15]
Flagellospora curvula	Declines in later samples [1, 13, 10, 14, 16, 18, 19]
	Increases in later samples [3]
Flagellospora penicillioides	Present throughout [20]
Flagellospora sp.	Early peak, then decline [12]
Geniculospora inflata	Present throughout [16]
Heliscus lugdunensis	Most common in early samples [2, 4]
	Most common in mid-decay [6]
	More common in later samples [1]
	Present throughout [5]
Lemonniera aquatica	Tends to increase in later samples [7]
	Tends to decrease in later samples [8, 12, 16, 19]
	Present throughout [17, 18, 20]
	No information [11]
Lemonniera pseudofloscula	No information [15]
Lemonniera terrestris	Early samples [9]
	Common throughout [16]
Leptocladia neglecta	Later samples [13]
Lunulospora curvula	Declines in later samples [4, 16]
Monotosporella tuberculata	Late samples [13]
Mycocentrospora angulata	Rapidly declines after early dominance [7]
	Common throughout except in early samples [8]

Table 3.1 (Continued)

Pyricularia aquatica	Increases at later samples [1]
Tetrachaetum elegans	Most common in early samples [9, 13, 16] Most common in mid-decay [8] Dominant in all but first sample [7] Throughout [13]
Tetracladium marchalianum	Absent from early samples, but later stable [1, 18] Declines in later samples [6, 14, 19] Present throughout [5, 17, 10, 12, 20] No information [11, 15]
Tricellula aquatica	Increased during later samples [12]
Tricladium angulatum	More common in early samples [5, 13]
Tricladium chaetocladium	Common throughout [4]
Tricladium splendens	Declines in later samples [8]

[a] *References*: [1] Abel [1985]; [2] Aimer [1989]; [3] Bärlocher [1982]; [4] Bärlocher [1991a]; [5] Bärlocher and Kendrick [1974]; [6] Bärlocher et al. [1978]; [7] Chamier and Dixon [1982]; [8] Chamier et al. [1984]; [9] Chauvet [1989]; [10] Chauvet et al. [1986]; [11] Chergui [1990]; [12] Chergui and Pattée [1988a]; [13] Gönczöl [1989]; [14] Rossi et al. [1983]; [15] Shearer and Lane [1983]; [16] Shearer and Webster [1985b]; [17] Suberkropp [1984]; [18] Suberkropp and Klug [1976]; [19] Suberkropp et al. [1988]; [20] van der Merwe and Jooste [1988].

Succession and Island Biogeography. During the decay of a substratum, different waves of fungi may become dominant. This process is referred to as a succession, and has received much attention from mycologists (Frankland 1981). Many factors have been demonstrated to influence the appearance of the various species; nevertheless, there is no clear understanding of the actual factors that allow the replacement of one species by another. Connell and Slatyer (1977) provided a good discussion of mechanisms that might influence succession. They differentiate between facilitation, tolerance, and inhibition models. These terms describe the effect of early colonizers on later arrivals and are largely self-explanatory. In reality, successions are likely to be messy, and contain elements of all three models. Strictly speaking, succession must involve replacement of mycelium of one species by another at the same location (Swift 1976). In fungal successions, in situ identifications are difficult, and destructive sampling must generally be used. The standard procedure is to introduce many replicate samples at the same time, and periodically remove a certain number of them. This inevitably introduces additional uncertainties.

In recent years, the similarity of dung pats, leaves and twigs to islands has been noted repeatedly. Therefore it was hoped that their colonization and succession might follow rules suggested by MacArthur and Wilson (1973). In their model, the number of species on islands tends towards a stable value, which reflects the balance between rates of immigration and extinction. It represents a dynamic equilibrium, therefore there should be evidence for turnover of species. At least in its original form, island biogeography ignores potential effects of competition on species occurrence (Gilbert 1980). It was later modified, to allow existence of a permanent core of resident species (Diamond and May 1976). There is one obvious difference between real islands and leaves or twigs: the latter disappear completely within a year or less due to decay. They have been compared to a slowly sinking archipelago (Bärlocher 1982). If applicable at all, the theory of island biogeography would only be valid during an early period. This may have contributed to the fact that there have been only half-hearted attempts to apply it to aquatic hyphomycetes.

Sanders and Anderson (1979) found that the number of fungal species increased with the size of wood blocks. Six species were found on blocks of all sizes, a further seven species were present on all but the smallest blocks and another six species were found occasionally on the largest blocks only. Sanders and Anderson (1979) concluded that the six species form a more or less stable association which excludes other species on the smallest blocks. As the size of the blocks increases, gaps may appear, which allow random colonization by less successful species. During the three months of the experiment there was no evidence for turnover of the main six species. Their results provide some support for the species/area relationship, which the theory of island biogeography tries to explain (Preston 1962; Gilbert 1980).

Chamier and Dixon (1982) reached somewhat similar conclusions based on a study with alder leaves: they found four dominant species and about ten occasional species. The average number of species per leaf pack was 14.

Bärlocher and Schweizer (1983) found a significant correlation between log (leaf area) and cumulative number of fungal colonizers. The goodness of fit of the regression equation was considerably improved when integrated decay curves instead of initial areas or weights of the squares were used, indicating that the length of substrate availability is correlated with the number of fungal species able to colonize it. At the later stages, the rate of arrival of new species slowed down, but did not fall to zero.

Looking at all studies published to date, there is a strong impression that an orderly sequence of individual fungi, or groups of fungi, becoming dominant at successive stages of leaf decay is rare. Similarities between communities of successive samples are generally high, though there may be "an element of succession" on slowly processed leaves (Chamier and Dixon 1982). The major exceptions are fungi responding to seasonal changes in water temperature (Webster et al. 1976; Suberkropp 1984). The fungal community on leaves is generally dominated by four to eight species that arrive early and persist throughout the substratum's existence. There are no obvious and consistent patterns suggesting that less competitive species are being displaced at later stages, nor that chemical changes result in a nutritional relay mechanism (Egler 1954; Suberkropp and Klug 1976; Rosset et al. 1982). It is as if the leaf receives an "imprint" upon its submersion in a stream; this imprint largely determines the fungal community (Bärlocher 1982). Mycelial growth is closely followed by sporulation; both increase to an early peak and later decline (Suberkropp 1991a,b). As a result, success (as measured by biomass, or more appropriately by reproductive output) of a species is closely correlated with its arrival time on the leaf. The importance of early arrival is emphasized by the rapid decay and deterioration of leaves by prior fungal colonization (Suberkropp 1984; Rosset and Bärlocher 1985).

Niche Parameters and Species Diversity. Spore size, surface and shape, selective feeding by invertebrates, susceptibility to leaf toxins, water chemistry and temperature, differences in enzymatic equipment, survival outside the stream, all potentially influence the success of an aquatic hyphomycete. With the exception of temperature, no single factor is generally strong enough to explain the absence or presence of a species in a given stream. Even if we assume that each of the nine factors enumerated above has only two levels, over 500 combinations could be defined, which exceeds the total number of aquatic hyphomycetes species described. If for each of these combinations a different species were dominant, one would expect to find more than the usual 20–60 species coexisting in a stream. On the other hand, one species might dominate at many more than one of these combinations. It is therefore not obvious whether one has to explain why so many, or, why so few, species actually coexist in a given stream. In laboratory experiments, however, I have found it impossible to routinely prepare individual leaves supporting 5–15 fungal species (unpubl. obs.). Such leaves are common in streams. When starting with leaves collected from a stream, and transferring spores produced from them to sterile leaves, most of the originally present species are unsuccessful. Within a few transfers, only one or two remain. This seems to indicate that on most naturally occurring leaves, no competitive equilibrium has been reached. Maybe the crucial ingredients absent from laboratory studies are the size and complexity of the natural system.

3.2.2.5 Ephemeral and Divided Resources

Aquatic hyphomycetes face a problem found in other groups: how to survive on a divided and ephemeral resource. Many invertebrates exploit such patchy resources: dung, carrion, fruits, seeds, etc. Shorrocks et al. (1979) proposed a model for fruit flies where inferior competitors rely on the random occurrence of refuges in which larvae of the best competitors are rare. High intraspecific aggregation, and random distribution of the species reduce interspecific competition and increase intraspecific competition (Shorrocks and Rosewell 1987). In the one study that subjected spatial distribution of fungi on leaves to statistical analysis, there was indeed a slight, if transitory, trend from initially random to a more clumped distribution (Chamier et al. 1984). Clearly, this aspect might repay further study.

The enormous size and complexity of the natural system, coupled with the periodicity of food supply, may prevent the aquatic hyphomycete community from ever reaching an equilibrium. Competition may periodically be severe (and this would clearly show up in bottle experiments), but its importance in determining community structure may well be exceeded by other factors. Maybe Hutchinson's remarks on phytoplankton (1967) apply to aquatic hyphomycetes as well: "The competitors of a given genus or other higher taxon are from time to time lined up, and sometimes the race begins, but as it might be in the works of Lewis Carroll, the event is always called off before it is completed and something entirely different is arranged in its place." The way to tackle these problems might be some application of Caswell's model (1978), consisting of many semi-separate cells and random exchange.

3.3 Conclusions

To return to the simile at the beginning of this Chapter: Baron von Münchhausen's solution was to shoot the bridle in two. Once again, his horse was on firm ground. Münchhausen jumped on his horse, and galloped away (von Baeyer 1986). There is considerable optimism among some physicists that they might yet reach their aim of explaining everything by a grand unified theory. The picture of a heroic scientist, sitting in his wheelchair, and figuring out the world's mysteries in his head, continues to have enormous appeal. Regardless of its application to physics, such an ideal is surely inappropriate for ecologists. Like economists, the best they can hope for are "stylized facts" (Kaldor 1985). I'm convinced that what is somewhat condescendingly labelled as scientific natural history has contributed more to an understanding of aquatic hyphomycetes, and will continue to do so, than mathematical theory. The primary aim should not be the possibility to instruct biologists working in other areas, but to learn more about the system at hand. Like other areas strongly influenced by historical events, it demands explanation in the form of a narrative in which specific events, not unchanging laws, have pride of place (Gould 1989).

References

Abdullah SK, Taj-Aldeen SJ (1989) Extracellular enzymatic activity of aquatic and aero-aquatic conidial fungi. Hydrobiologia 174:217–223

Abel T (1985) Einfluss von Cadmium auf aquatische Hyphymyzeten und deren Konsument *Gammarus fossarum* Koch. Thesis, University of Basel, Switzerland

Aimer RD (1989) Ecology of aquatic hyphomycetes in New Zealand streams. Thesis, University of Waikato, New Zealand

Aimer RD, Segedin BP (1985) Some aquatic hyphomycetes from New Zealand streams. NZ J Bot 23:273–299

Akridge RE, Koehn TD (1987) Amphibious hyphomycetes from the San Marcos River in Texas. Mycologia 79:228–233

Andrewartha HG, Birch LC (1954) Distribution and abundance of animals. University of Chicago Press, Chicago

Arsuffi TL, Suberkropp K (1985) Selective feeding by stream caddisfly (Trichoptera) detritivores on leaves with fungal-colonized patches. Oikos 45:50–58
Bärlocher F (1979) On trophic interactions between microorganisms and animals. Am Nat 113:147–148
Bärlocher F (1980) Leaf-eating invertebrates as competitors of aquatic hyphomycetes. Oecologia 47:303–306
Bärlocher F (1981) Fungi on the food and in the faeces of *Gammarus pulex*. Trans Br Mycol Soc 76:160–165
Bärlocher F (1982) Conidium production from leaves and needles in four streams. Can J Bot 60:1487–1494
Bärlocher F (1985) The role of fungi in the nutrition of stream invertebrates. Bot J Linn Soc 91:83–94
Bärlocher F (1990) Factors that delay colonization of fresh alder leaves by aquatic hyphomycetes. Arch Hydrobiol 119:249–255
Bärlocher F (1991a) Fungal colonization of fresh and dried leaves in the River Teign. Nova Hedwigia 52:349–357
Bärlocher F (1991b) Intraspecific hyphal interactions among aquatic hyphomycetes. Mycologia 83:82–88
Bärlocher F, Kendrick B (1973) Fungi and food preferences of *Gammarus pseudolimnaeus*. Arch Hydrobiol 72:501–516
Bärlocher F, Kendrick B (1974) Dynamics of the fungal population on leaves in a stream. J Ecol 63:761–791
Bärlocher F, Oertli JJ (1978) Inhibitors of aquatic hyphomycetes in dead conifer needles. Mycologia 70:964–974
Bärlocher F, Rosset J (1981) Aquatic hyphomycete spora of two Black Forest and two Swiss Jura streams. Trans Br Mycol Soc 76:479–483
Bärlocher F, Schweizer M (1983) Effects of leaf size and decay rate on colonization by aquatic hyphomycetes. Oikos 41:205–210
Bärlocher F, Kendrick B, Michaelides J (1977) Colonization of rosin-coated slides by aquatic hyphomycetes. Can J Bot 55:1163–1166
Bärlocher F, Kendrick B, Michaelides J (1978) Colonization and conditioning of *Pinus resinosa* needles by aquatic hyphomycetes. Arch Hydrobiol 81:462–474
Barnes JR, Minshall GW (1981) Stream ecology. Application and testing of general ecological theory. Plenum, New York
Bengtsson G (1983) Habitat selection in two species of aquatic hyphomycetes. Microb Ecol 9:15–26
Betancourt C, Cruz J, Garcia J (1987) Los hifomicetos acuaticos de la quebrada Dona Juana en el bosque estatal de Toro Negro, Villalba, Puerto Rico. Caribb J Sci 23:278–284
Brown, JH, Davidson DW (1977) Competition between seed-eating rodents and ants in desert ecosystems. Science 196:880–882
Caswell H (1978) Predator-mediated coexistence: a nonequilibrium model. Am Nat 112:127–154
Chamier A-C (1985) Cell-wall degrading enzymes of aquatic hyphomycetes: a review. Bot J Linn Soc 91:67–81
Chamier A-C, Dixon PA (1982) Pectinases in leaf degradation by aquatic hyphomycetes. I: the field study. Oecologia 52:109–115
Chamier A-C, Dixon PA, Archer SA (1984) The spatial distribution of fungi on decomposing alder leaves in a freshwater stream. Oecologia 64:92–103
Chandrashekar KR, Kaveriappa KM (1988) Production of extracellular enzymes by aquatic hyphomycetes. Folia Microbiol 33:55–58
Chandrashekar KR, Sridhar KR, Kaveriappa KM (1990) Periodicity of water-borne hyphomycetes in two streams of Western Ghat forests (India). Acta Hydrochim Hydrobiol 18:187–204
Chauvet E (1989) Production, flux et décomposition des litières en milieu alluvial
Dynamique et rôle des hyphomycètes aquatiques dans le processus de décomposition. Thesis, Université de Toulouse, France

Chauvet E, Mercé J (1988) Hyphomycètes aquatiques: importance dans la décomposition des litières. Rev Sci Eau 1:203–216
Chauvet E, Mercé J, Jean-Louis A-M (1986) Hyphomycètes aquatiques colonisant les feuilles de *Salix alba* L. Bull Soc Mycol Fr 102:347–351
Chergui H (1990) The dynamics of aquatic hyphomycetes in an eastern Moroccon stream. Arch Hydrobiol 118:341–352
Chergui H, Pattée E (1988a) The dynamics of Hyphomycetes on decaying leaves in the network of the River Rhone (France). Arch Hydrobiol 114:3–20
Chergui H, Pattée E (1988b) The impact of benthic invertebrates on the breakdown of poplar leaves in the network of a large European river. Arch Hydrobiol 113:15–25
Christensen M (1989) A view of fungal ecology. Mycologia 81:1–19
Cohen JE (1971) Mathematics as metaphor. Science 172:674–675
Collier KJ (1990) Ingestion of sooty mould fungi by some New Zealand stream insects. N Z Entomol 13:42–45
Colwell RK, Winkler DW (1984) A null model for null models in biogeography. In: Strong DR, Simberloff D, Abele LG, Thistle AB (eds) Ecological communities. Conceptual issues and the evidence. Princeton University Press, Princeton, pp 344–359
Connell JH (1975) Producing structure in natural communities. In: Cody ML, Diamond JM (eds) Ecology and evolution of communities. Harvard University Press, Cambridge, pp 460–490
Connell JH (1978) Diversity in tropical rain forests and coral reefs. Science 199:1302–1310
Connell JH (1980) Diversity and coevolution of competitors, or the ghost of competition past. Oikos 35:131–138
Connell JH, Slatyer RO (1977) Mechanisms of succession in natural communities and their role in community stability and organisation. Am Nat 111:1119–1144
Conway KE (1970) The aquatic hyphomycetes of central New York. Mycologia 62:516–530
Cooke RC, Rayner ADM (1984) Ecology of saprotrophic fungi. Longman, London
Covich AP (1988) Geographical and historical comparisons of neotropical streams: biotic diversity and detrital processing in highly variable habitats. J N Am Benthol Soc 7:361–386
Cummins KW (1974) Structure and function of stream ecosystems. BioScience 245:631–641
Darwin C (1859) On the origin of species. Murray, London
Dayton PK (1980) Citation classic. Curr Cont 11:18
Del Frate G, Caretta G (1983) Aquatic hyphomycetes of a mountain stream in Valesia (Piemonte). Hydrobiologia 102:69–71
Diamond JM, May RM (1976) Island biogeography and the design of natural reserves. In: May RM (ed) Theoretical ecology. Sinauer, Sunderland, Massachusetts, pp 228–252
Egerton FN (1973) Changing concepts of the balance of nature. Q Rev Biol 48:322–350
Egler FE (1954) Vegetation science concepts. I. Initial floristic composition – a factor in old-field vegetation development. Vegetatio 4:412–417
Elton C (1927) Animal ecology. Clowes and Sons, London
Elton C (1930) Animal ecology and evolution. Oxford University Press, Oxford
Elton C (1933) The ecology of animals. Reprinted 1966 by Science Paperbacks and Methuen, London
Ferreira JC, Sinclair RC, Eicker A (1981) The aquatic hyphomycetes of the Apies River, South Africa. S Afr J Bot 47:49–55
Findlay SEG, Arsuffi TL (1989) Microbial growth and detritus transformations during decomposition of leaf litter in a stream. Freshwater Biol 21:261–269
Fisher PJ, Webster J (1981) Ecological studies of aero-aquatic hyphomycetes. In: Wicklow DT, GC Carroll (eds) The fungal community. Marcel Dekker, New York, pp 709–730
Fisher SG (1977) Organic matter processing by a stream segment ecosystem: Fort River, Massachusetts, USA. Int Rev Gesamten Hydrobiol 62:701–727

Frankland JC (1981) Mechanisms in fungal successions. In: Wicklow DT, Carroll GC (eds) The fungal community. Marcel Dekker, New York, pp 403–426
Garrett SD (1952) The soil fungi as a microcosm for ecologists. Sci Progr 40:436–450
Garrett SD (1963) Soil fungi and soil fertility. Pergamon Press, New York
Gause GF (1934) The struggle for existence. Williams and Wilkins, Baltimore (reprinted 1971 by Dover, New York)
Gessner MO, Schwoerbel J (1989) Leaching kinetics of fresh leaf-litter with implications for the current concept of leaf-processing in streams. Arch Hydrobiol 115:81–90
Gilbert FS (1980) The equilibrium theory of island biogeography: fact or fiction? J Biogeogr 7:209–235
Giovannoni SJ, Britschgi TB, Moyer CL, Field KG (1990) Genetic diversity in Sargasso Sea bacterioplankton. Nature 345:60–62
Gleason H (1926) The individualistic concept of the plant association. Bull Torrey Bot Club 53:1–20
Godfrey BES (1983) Growth of two terrestrial microfungi on submerged alder leaves. Trans Br Mycol Soc 81:418–421
Gönczöl J (1971) Aquatic hyphomycetes from the Mts. Börzsöny. Ann Hist Nat Mus Natl Hung 63:57–75
Gönczöl J (1975) Ecological observations on the aquatic hyphomycetes of Hungary. Acta Bot Acad Sci Hung 21:243–264
Gönczöl J (1987) Ecological observations on aquatic hyphomycetes of Hungary, III. Acta Bot Hung 33:41–49
Gönczöl J (1989) Longitudinal distribution patterns of aquatic hyphomycetes in a mountain stream in Hungary. Experiments with leaf packs. Nova Hedwigia 48:391–404
Goos RD (1987) Fungi with a twist: the helicosporous Hyphomycetes. Mycologia 79:1–22
Gould SJ (1989) Wonderful life. Hutchinson Radius, London
Greathead SK (1961) Some aquatic hyphomycetes in South Africa. J S Afr Bot 27:195–228
Gunasekera SA, Webster J (1983) Inhibitors of aquatic and aero-aquatic hyphomycetes in pine and oak wood. Trans Br Mycol Soc 80:121–125
Haeckel J-W, Meijering MPD, Rusetzki H (1973) *Gammarus fossarum* Koch als Fallaubzersetzer in Waldbächen. Freshwater Biol 3:241–249
Hairston, NG, Smith FE, Slobodkin LB (1960) Community structure, population control, and competition. Am Nat 94:421–425
Hardin G (1960) The competitive exclusion principle. Science 131:1292–1297
Hutchinson GE (1957) Concluding remarks. Cold Spring Harbor Symp Quant Biol 22:415–427
Hutchinson GE (1967) A treatise on limnology vol II: Introduction to lake biology and limnoplankton. John Wiley, New York
Hutchinson GE (1975) Variations on a theme by Robert MacArthur. In: Cody ML, Diamond JM (eds) Ecology and evolution of communities. Harvard University Press, Cambridge
Ingold CT (1942) Aquatic hyphomycetes of decaying alder leaves. Trans Br Mycol Soc 25:339–417
Ingold CT (1975) An illustrated guide to aquatic and waterborne hyphomycetes (Fungi Imperfecti) with notes on their biology. Freshwater Biol Assoc Sci Publ No 30
Iqbal SH, Bhatty SF (1979) Conidia from stream foam. Trans Mycol Soc Jpn 20:83–91
Iqbal SH, Webster J (1973a) Aquatic hyphomycete spora of the River Exe and its tributaries. Trans Br Mycol Soc 61:331–346
Iqbal SH, Webster J (1973b) The trapping of aquatic hyphomycete spora by air bubbles. Trans Br Mycol Soc 60:37–48
Iqbal SH, Webster J (1977) Aquatic hyphomycete spora of some Dartmoor streams. Trans Br Mycol Soc 69:233–241
Iqbal SH, Bhatty SF, Malik KS (1979) Fresh-water hyphomycetes on decaying plant debris submerged in some streams of Pakistan. Trans Mycol Soc Jpn 20:51–61

Iqbal SH, Bhatty SF, Malik KS (1980) Freshwater hyphomycetes on submerged decaying pine needles in Pakistan. Trans Mycol Soc Jpn 21:321–327

Iversen TM (1973) Decomposition of autumn-shed beech leaves in a springbrook and its significance for the fauna. Arch Hydrobiol 72:305–312

Kaldor N (1985) Economics without equilibrium. Sharpe, Armonk, New York

Kaushik NK, Hynes HBN (1971) The fate of the dead leaves that fall into streams. Arch Hydrobiol 68:465–515

Kendrick B (1979) The whole fungus vols 1 and 2. National Museum of Natural Sciences, Ottawa

Khan MA (1987) Interspecies interactions in aquatic hyphomycetes. Bot Mag Tokyo 100:295–303

Kirby JJH, Webster J, Baker JH (1990) A particle plating method for analysis of fungal community composition and structure. Mycol Res 94:621–626

Koske RE, Duncan IW (1974) Temperature effects on growth, sporulation and germination of some "aquatic" hyphomycetes. Can J Bot 52:1387–1391

Lamore BJ, Goos RD (1978) Wood-inhabiting fungi of a freshwater stream in Rhode Island. Mycologia 70:1025–1034

Lindsey BI, Glover BJ (1976) Ecological studies of spores of aquatic Hyphomycetes in the Cringle Brook, Lincs. Hydrobiologia 51:201–208

MacArthur RH (1972) Geographical Ecology. Princeton University Press, Princeton

MacArthur RH, Wilson EO (1973) The theory of island biogeography. Princeton University Press, Princeton

May RM (1974) Scaling in ecology. Science 184:682–683

May RM (1976) Simple mathematical models with very complicated dynamics. Nature 261:459–467

May R (1989) The chaotic rhythms of life. New Scientist, Nov 18:37–41

McIntosh RP (1985) The background of ecology. Cambridge University Press, Cambridge

Mer GS, Khulbe RD (1981) Aquatic hyphomycetes of Kumaun Himalaya, India. Sydowia 34:118–124

Mer GS, Sati SC (1989) Seasonal fluctuation in species composition of aquatic hyphomycetous flora in a temperate freshwater stream of Central Himalaya, India. Int Rev Ges Hydrobiol 74:433–437

Mercé J (1987) Hyphomycètes aquatiques. Etude des variations saisonnières d'une population. Cryptogam Mycol 8:1–11

Metvalli AA, Shearer CA (1989) Aquatic hyphomycete communities in clear-cut and wooded areas of an Illinois stream. Trans Ill Acad Sci 82:5–16

Michaelides J, Kendrick B (1978) An investigation of factors retarding colonization of conifer needles by amphibious hyphomycetes in streams. Mycologia 70:419–430

Mohamad M, Keudell KC, Gessner RV (1989) Use of ELISA to differentiate marine species of *Dendryphiella*. Mycol Res 93:400–402

Müller-Haeckel A (1977) Annual production of Hyphomycete conidia in a subarctic stream. Oikos 29:396–397

Müller-Haeckel A, Marvanová L (1976) Konidienproduktion und -kolonisation von Süsswasser-Hyphomyzeten im Kaltisjokk (Lappland). Bot Not 129:405–409

Müller-Haeckel A, Marvanová L (1979) Periodicity of aquatic hyphomycetes in the subarctic. Trans Br Mycol Soc 73:109–116

Müller-Haeckel A, Rheinheimer G (1983) Studies on the annual cycle of bacteria and fungi in the Ängerån, a coastal stream in northern Sweden. Aquilo Ser Zool 22:51–56

Myers J, Smith J, Krebs C (1990) Chaos rules. New Scientist 31 (March):72

Newell SY (1991) Estimating fungal biomass and productivity. In: Carroll GC, Wicklow DT (eds) The fungal community, 2nd edn. Marcel Dekker, New York (in press)

Newton JA (1971) A mycological study of decay of the leaves of deciduous trees on the bed of a river. Thesis, University of Salford, England

Nilsson S (1964) Freshwater hyphomycetes: taxonomy, morphology and ecology. Symb Bot Ups 18:1–130

Paine R (1966) Food web complexity and species diversity. Am Nat 100:65–75

Petersen RC, Cummins KW (1974) Leaf processing in a woodland stream. Freshwater Biol 4:343–368

Pielou EC (1969) An Introduction to Mathematical Ecology. Wiley, New York

Preston FW (1962) The canonical distribution of commonness and rarity. Ecology 43:185–215, 410–432

Prus T (1981) Decomposition of leaf litter by *Asellus aquaticus L.* (Isopoda, Crustacea). Verh Int Ver Limnol 21:1584–1588

Pugh GJF (1980) Strategies in fungal ecology. Trans Br Mycol Soc 75:1–14

Puppi G (1983) Occurrence of filamentous fungi on decaying leaves in lake waters (Albano and Nemi, central Italy). Ann Bot 41:27–36

Ranzoni FV (1953) The aquatic hyphomycetes of California. Farlowia 4:353–398

Ranzoni FV (1979) Aquatic hyphomycetes from Hawaii. Mycologia 71:786–795

Rayner ADM, Franks NR (1987) Evolutionary and ecological parallels between ants and fungi. TREE 2:127–133

Rayner ADM, Todd NK (1982) Population structure in wood-decomposing basidiomycetes. In: Frankland JC, Hedger JN, Swift MJ (eds) Decomposer basidiomycetes. Cambridge University Press, Cambridge, pp 108–128

Root RB (1967) The niche exploitation pattern of the blue-gray gnatcatcher. Ecol Monogr 37:317–350

Rosset J, Bärlocher F (1985) Transplant experiments with aquatic hyphomycetes. Verh Int Ver Limnol 22:2786–2790

Rosset J, Bärlocher F, Oertli JJ (1982) Decomposition of conifer needles and deciduous leaves in two Black Forest and two Swiss Jura streams. Int Rev Ges Hydrobiol 67: 695–711

Rossi L, Fano EA, Basset A, Fanelli C, Fabbri AA (1983) An experimental study of a microfungal community on plant detritus in a mediterranean woodland stream. Mycologia 75:887–896

Roughgarden J (1989) The structure and assembly of communities. In: Roughgarden J, May RM, Levin SA (eds) Perspectives in ecological theory. Princeton University Press, Princeton NJ, pp 203–226

Salt GW (1979) A comment on the term emergent properties. Am Nat 113:145–148

Sanders PF, Anderson JM (1979) Colonization of wood blocks by aquatic hyphomycetes. Trans Br Mycol Soc 73:103–107

Schoener TW (1972) Mathematical ecology and its place among the sciences. Science 178:389–391

Scudo F, Ziegler JR (1978) The golden age of theoretical ecology, 1923–1940. Springer, Berlin Heidelberg New York

Shearer CA, Lane L (1983) Comparison of three techniques for the study of aquatic hyphomycete communities. Mycologia 75:498–508

Shearer CA, Webster J (1985a) Aquatic hyphomycete communities in the River Teign. I. Longitudinal distribution patterns. Trans Br Mycol Soc 84:489–501

Shearer CA, Webster J (1985b) Aquatic hyphomycete communities in the River Teign. II. Temporal distribution patterns. Trans Br Mycol Soc 84:503–507

Shearer CA, Webster J (1985c) Aquatic hyphomycete communities in the River Teign. III. Comparison of sampling techniques. Trans Br Mycol Soc 84:509–518

Shearer CA, Zare-Maivan H (1988) In vitro hyphal interactions among wood- and leaf-inhabiting Ascomycetes and Fungi Imperfecti from freshwater habitats. Mycologia 80:31–37

Shorrocks B, Rosewell J (1987) Spatial patchiness and community structure: coexistence and guild size of *Drosophila* on ephemeral resources. In: Gee JHR, Giller PS (eds) The Organisation of Communities: Past and Present, Blackwells, Oxford, pp 29–51

Shorrocks B, Atkinson W, Charlesworth P (1979) Competition on a divided and ephemeral resource. J Anim Ecol 48:899–908

Simberloff D (1980) A succession of paradigms in ecology: essentialism to materialism and probabilism. Synthese 43:3–39

Simberloff D (1982) The status of competition theory in ecology. Ann Zool Fenn 19:241–253

Sinclair RC, Ebersohn C, Eicker A (1983) The aquatic hyphomycetes of the Hennops River (Irene), South Africa. S Afr J Bot 2:224–230

Singh N, Musa TM (1977) Terrestrial occurrence and the effect of temperature on growth, sporulation and spore germination of some tropical aquatic hyphomycetes. Trans Br Mycol Soc 68:103–106

Sridhar KR, Kaveriappa KM (1984) Seasonal occurrence of water-borne fungi in Konaje stream (Mangalore), India. Hydrobiologia 119:101–105

Sridhar KR, Kaveriappa KM (1987) Culturing water-borne hyphomycetes on plant latex. J Indian Bot Soc 66:232–233

Sridhar KR, Kaveriappa KM (1988) Colonization of leaf litter by aquatic hyphomycetes in a tropical stream. Arch Hydrobiol 112:627–630

Sridhar KR, Kaveriappa KM (1989) Colonization of water-borne hyphomycetes in a tropical stream. Mycol Res 92:392–396

Stewart I (1989) Does God play dice? Penguin, London

Strong DR (1986) Density vagueness: the variance in the demography of real populations. In: Diamond JM, Case TJ (eds) Community ecology, Harper and Row, New York, pp 257–268

Strong DR, Simberloff D, Abele LG, Thistle AB (1984) Ecological communities. Conceptual issues and the evidence. Princeton University Press, Princeton

Suberkropp K (1984) Effect of temperature on seasonal occurrence of aquatic hyphomycetes. Trans Br Mycol Soc 82:53–62

Suberkropp K (1991a) Aquatic hyphomycete communities. In: Carroll GC, Wicklow DT (eds) The fungal community, 2nd edn. Marcel Dekker, New York (in press)

Suberkropp K (1991b) Relationships between growth and sporulation of aquatic hyphomycetes on decomposing leaf litter. Mycol Res 95:843–850

Suberkropp K, Klug MJ (1974) Decomposition of deciduous leaf litter in a woodland stream. I. A scanning electron microscopic study. Microb Ecol 1:96–103

Suberkropp K, Klug MJ (1976) Fungi and bacteria associated with leaves during processing in a woodland stream. Ecology 57:707–719

Suberkropp K, Klug MJ (1980) The maceration of deciduous leaf litter by aquatic hyphomycetes. Can J Bot 58:1025–1031

Suberkropp K, Klug MJ (1981) Degradation of leaf litter by aquatic hyphomycetes. In: Suberkropp K, Godshalk GL, Klug MJ (1976) Changes in the chemical composition of leaves during processing in a woodland stream. Ecology 57:720–727

Suberkropp K, Arsuffi TL, Anderson JP (1983) Comparison of degradative ability, enzymatic activity, and palatability of aquatic hyphomycetes grown on leaf litter. Appl Environ Microbiol 46:237–244

Suberkropp K, Michelis A, Lorch H-J, Ottow JCG (1988) Effect of sewage treatment plant effluents on the distribution of aquatic hyphomycetes in the River Erms, Schwäbische Alp, F.R.G. Aquat Bot 32:141–153

Sugihara G, May RM (1990) Nonlinear forecasting as a way of distinguishing chaos from measurement error in time series. Nature 344:734–741

Summerbell R, Cannings RA (1981) Fungi and *Brillia retifinis* (Diptera: Chironomidae) in the decomposition of conifer leaves in a British Columbia stream. Syesis 14:125–130

Swift MJ (1976) Species diversity and the structure of microbial communities. In: Anderson JM, Macfadyen A (eds) The role of aquatic and terrestrial organisms in decomposition processes. Blackwell, Oxford, pp 185–222

Thomas K, Chilvers GA, Norris RH (1989) Seasonal occurrence of conidia of aquatic hyphomycetes (Fungi) in Lees Creek, Australian Capital Territory. Aust J Mar Freshwater Res 40:11–23

Thompson W, Rayner ADM (1982) Spatial structure of a population of *Tricholomopsis platyphylla* in a woodland site. New Phytol 92:103–114

Townsend CR, Hildrew AG, Schofield K (1987) Persistence of stream invertebrate communities in relation to environmental variability. J Anim Ecol 56:597–614

Triska FJ (1970) Seasonal distribution of aquatic hyphomycetes in relation to the disappearance of leaf litter from a woodland stream. Thesis, University of Pittsburgh, Pittsburgh

Tubaki K, Tokumasu S, Ando K (1985) Morning dew and *Tripospermum* (Hyphomycetes). Bot J Linn Soc 91:45–50

Van der Merwe WJJ, Jooste WJ (1988) A synecological study of aquatic Hyphomycetes in the Mooi river, western Transvaal, and their significance in the decomposition of allochthonous leaf litter. S Afr J Sci 84:314–320

Von Baeyer HC (1986) The odds against order. The Sciences, March/April, pp 2–4

Walker I (1985) On the structure and ecology of the micro-fauna in the Central Amazonian forest stream "Igarapé da Cachoeira". Hydrobiologia 122:137–152

Warcup JH (1957) Studies on the occurrence and activity of fungi in a wheat field soil. Trans Br Mycol Soc 40:237–259

Waterman TH (1965) The problem. In: Waterman TH, Morowitz HJ (eds) Theoretical and mathematical biology. Blaisdell, New York, pp 3–23

Webster J (1975) Further studies of sporulation of aquatic hyphomycetes in relation to aeration. Trans Br Mycol Soc 64:119–127

Webster JR, Benfield EF (1986) Vascular plant breakdown in freshwater ecosystems. Annu Rev Ecol Syst 17:567–594

Webster J, Descals E (1981) Morphology, distribution and ecology of conidial fungi in freshwater habitats. In: Cole GT, Kendrick B (eds) Biology of Conidial Fungi, vol 1. Academic Press, New York, pp 295–355

Webster J, Moran ST, Davey RA (1976) Growth and sporulation of *Tricladium chaetocladium* and *Lunulospora curvula* in relation to temperature. Trans Br Mycol Soc 67:491–549

Weldon CW, Slauson WL (1986) The intensity of competition versus its importance: an overlooked distinction and some implications. Q Rev Biol 61:23–44

Wicklow DT, Carroll GC (1981) Introduction. In: Wicklow DT, Carroll GC (eds) The fungal community. Marcel Dekker, New York, pp 1–8

Wiens JA (1977) On competition and variable environments. Am Sci 65:590–597

Willoughby LG, Archer JF (1973) The fungal spora of a freshwater stream and its colonization pattern on wood. Freshwater Biol 3:219–239

Willoughby LG, Minshall GW (1975) Further observations on *Tricladium giganteum*. Trans Br Mycol Soc 65:77–82

Wilson EO (1969) The new population biology. Science 163:1184–1185

Wood-Eggenschwiler S, Bärlocher F (1983) Aquatic hyphomycetes in 16 streams in France, Germany and Switzerland. Trans Br Mycol Soc 81:371–379

Yodzis P (1989) Introduction to theoretical ecology. Harper and Row, New York

Zare-Maivan H, Shearer CA (1988) Wood decay activity and cellulase production by freshwater lignicolous fungi. Int Biodeterior Bull 24:459–474

4 The Role of Woody Debris

C.A. Shearer

4.1 Introduction

Significant quantities of woody debris are found in stream channels draining forested watersheds. For example, as much as 250–800 tons/hectare of woody debris were found in stream channels of old growth conifer forests compared to 40–130 t/ha in streams draining mixed hardwood forests (Triska and Cromack 1980). Once in a stream, woody debris has a much longer residence time than leaves due to its slow rate of decomposition and its large size which resists displacement (Swanson et al. 1976). In stream systems, woody debris influences channel morphology, increases habitat diversity, affects transport of other materials by forming debris dams and serves as a carbon source, site of attachment and shelter for aquatic organisms (Fisher and Likens 1972; Anderson and Sedell 1979; Bilby and Likens 1980; Bird and Kaushik 1981; Harmon et al. 1986).

Despite its abundance and importance in rivers and streams, little is known about how woody debris is processed. The white and brown rot fungi and xylophagous insects that are important decomposers of wood in terrestrial habitats are largely absent in aquatic habitats (Harmon et al. 1986). Although fungi occur commonly and abundantly on submerged wood (Jones and Oliver 1964; Shearer 1972; Willoughby and Archer 1973; Lamore and Goos 1978; Shearer and Crane 1978; Sanders and Anderson 1979; Shearer and von Bodman 1983; Shearer and Crane 1986; Révay and Gönczöl 1990; Shearer and Webster 1991) ecologists have been slow to appreciate their possible role. A common impression exists that bacteria and actinomycetes are more important than fungi in processing submerged wood (Aumen et al. 1983; Benner et al. 1986; Harmon et al. 1986). Reasons for the possible underestimation of the role of fungi in wood processing may be due to methodologies that favour bacteria (Benner et al. 1986) and the general lack of study of freshwater lignicolous fungi.

Although distributional and ecological studies of aquatic hyphomycetes on leaves are numerous and cover a wide geographical area (Webster and Descals 1981), only four studies of lignicolous aquatic hyphomycete communities exist (Willoughby and Archer 1973; Sanders and Anderson 1979;

Table 4.1. Comparison of wood and leaves[a]

Characteristics	Wood	Leaves
A. Physical-chemical features		
C:N ratio	High	Low
Pectins	Low	High
Parenchyma: sclerenchyma ratio	Low	High
Density	High	Low
Surface: volume ratio	Low	High
Size	Varied	Uniform
Input	Episodic	Seasonal
Residence time	Long	Short
B. Fungal Community Dynamics		
Initial colonization	Deterministic	Stochastic
Colonization pattern	Successive	Parallel
Successive communities	Several	Few
Functional group diversity	High?	Low?
Taxonomic diversity	High?	Low?
Competition	High	Low
Interdependence	High?	Low?
Teleomorph: anamorph ratio	High	Low
Grazer/fungal interactions	Low?	High
Rate of decomposition	Slow	Fast

[a] Characteristics vary among both wood and leaf species; the above comparisons are based on generalized relative differences between wood and leaves.

Révay and Gönczöl 1990; Shearer and Webster 1991). Possible reasons for this substratum bias include the difficulty in working with woody debris due to its size and low surface to volume ratio, the longer time required for such studies due to the slow decomposition rate of wood, and the difficulty in obtaining quantitative data other than frequency of occurrence. A number of physical-chemical characteristics differ between wood and leaves (Table 4.1A) and these characteristics influence the respective roles of these substrata and the community dynamics of heterotrophs associated with them (Table 4.1B). Table 4.1 presents hypotheses regarding the role of woody debris in the life cycles of aquatic hyphomycetes and will be discussed herein.

4.2 Occurrence of Aquatic Hyphomycetes on Wood

4.2.1 Sampling Techniques

Techniques for sampling aquatic hyphomycetes on wood usually involve the random gathering of wood and/or baiting with natural twigs or cut wood (Willoughby and Archer 1973; Shearer and von Bodman 1983; Révay and Gönczöl 1990; Shearer and Webster 1991). After collection, wood may be examined with a dissecting microscope directly or after incubation in water and spores scraped from the surface for identification. Washings of twigs or macerated suspensions of wood may be plated and resulting colonies examined. After initial examination, woody debris is often incubated in moist chambers to enhance development of teleomorphs and non-Ingoldian hyphomycetes (Shearer and von Bodman 1983). Recently, the technique of placing twigs in water in individual aeration chambers and then filtering the water through membranes, has been used to quantify aquatic hyphomycete conidial production on twigs (Shearer and Webster 1991).

4.2.2 Species Occurring on Wood

Eighty-six species in 46 anamorph genera have been reported from wood (Table 4.2). This represents slightly less than one third of the total number of described aquatic hyphomycete species. Anamorph genera with the greatest number of representatives on wood are *Tricladium*, *Anguillospora* and *Dendrospora*. Generally, in a given habitat fewer species occur on wood than are found on leaves or in water or foam (Willoughby and Archer 1973; Sanders and Anderson 1979; Révay and Gönczöl 1990; Shearer and Webster 1991). This suggests that wood selects only certain species. For studies in which fungal occurrences on wood were quantified, relatively few species (6–12) occurred frequently (Table 4.3). Of these species, only *Anguillospora longissima*, *Clavariopsis aquatica*, *Alatospora acuminata* and *Tricladium splendens* are common to three or more studies.

Aquatic hyphomycete species composition on wood differs along longitudinal gradients in a single river system (Révay and Gönczöl 1990; Shearer and Webster 1991) and community structure parallels that on leaves except that fewer species usually occur on wood than leaves (Shearer and Webster 1991). Species composition also varies temporally on wood (Willoughby and Archer 1973; Shearer and Webster 1991). In the River Teign, dominant species declined and were replaced over time by other species which assumed dominance (Shearer and Webster 1991). Such distinct species replacements are not usually observed on leaves; rather, initial colonizers persist as new species are added and then members of the community disappear as the substratum is degraded (Chamier and Dixon 1982; Bärlocher and Schweizer 1983). The longer persistence of wood compared to leaves and the gradual release of new substrata through the sloughing off of old colonized substrata may permit successive colonization of wood. Whether the species replacements observed by Shearer and Webster (1991) are truly successional or whether they are cyclical and related to species seasonality is not known. Data from several studies (Willoughby and Archer 1973; Shearer and von Bodman 1983; Révay and Gönczöl 1990) suggest that *Nectria* species and their associated anamorphs, such as *N. lugdunensis* and its anamorph *Heliscus lugdunensis*, are early successional species that

Table 4.2. Aquatic hyphomycetes reported from submerged wood in freshwater habitats

Species	References
Actinospora megalospora Ingold	Ingold (1952)
Alatospora acuminata Ingold	Jones and Oliver (1964); Willoughby and Archer (1973); Sanders and Anderson (1979); Kane (1980); Shearer and von Bodman (1983); Révay and Gönczöl (1990); Shearer and Webster (1991); Shearer and Bartolata (unpublished)[a]
Alatospora flagellata (Gönczöl) Marv.	Révay and Gönczöl (1990)
Anavirga dendromorpha Descals and Sutton	Descals and Sutton (1976); Hamad and Webster (1987); Révay and Gönczöl (1990); Shearer unpublished – Hemlock and Douglas Fir wood submerged in Berry Creek, Oregon
Anguillospora crassa Ingold	Webster (1961); Willoughby and Archer (1973); Lamore and Goos (1978); Kane (1980); Révay and Gönczöl (1990); Shearer and Webster (1991); Shearer unpublished – Alder wood submerged in Berry Creek, Oregon
Anguillospora furtiva Descals	Révay and Gönczöl (1990)
Anguillospora gigantea Ranzoni	Ranzoni (1953)
Anguillospora longissima (Sacc. and Syd.) Ingold	Sládečková (1963); Eaton and Jones (1971); Willoughby and Archer (1973); Lamore and Goos (1978); Sanders and Anderson (1979); Kane (1980); Révay and Gönczöl (1990); Shearer and Webster (1991); Shearer and Bartolata (unpublished)
Anguillospora pseudolongissima Ranzoni	Lamore and Goos (1978)
Anguillospora sp.	Jones and Oliver (1964); Shearer and von Bodman (1983); Révay and Gönczöl (1990)
Articulospora tetracladia Ingold	Willoughby and Archer (1973); Sanders and Anderson (1979); Shearer and Webster (1991)
Bacillispora inflata Iqbal and Bhatty	Iqbal and Bhatty (1980)
Brachiosphaeria jamaicensis (Crane and Dumont) Nawawi	Crane and Dumont (1975)[b]
Campylospora chaetocladia Ranzoni	Nilsson (1964); Kane (1980)
Campylospora filicladia Nawawi	Nawawi (1974)
Casaresia sphagnorum Fragoso	Perrott (1960) – (as *Ankistrocladium fuscum* Perrott); Lamore and Goos (1978); Sanders and Anderson (1979); Shearer and Webster (1991)
Clavariopsis aquatica de Wildeman	Sládečková (1963); Jones and Oliver (1964); Willoughby and Archer (1973); Révay and Gönczöl (1990); Shearer and Webster (1991); Shearer and Bartolata (unpublished)
Clavatospora longibrachiata (Ingold) Nilsson ex Marv. and Nilsson	Sládečková (1963); Sanders and Anderson (1979)

Table 4.2 (Continued)

Species	References
Condylospora flexuosa Nawawi and Kuthubutheen	Nawawi and Kuthubutheen (1988)
Condylospora gigantea Nawawi and Kuthubutheen	Nawawi and Kuthubutheen (1988)
Condylospora spumigena Nawawi	Nawawi and Kuthubutheen (1988)
Culicidospora aquatica Petersen	Petersen (1960); Shearer and Webster (1991)
Cylindrocarpon aquaticum (Nilsson) Marv. and Descals	Shearer and von Bodman (1983)
Dendrospora erecta Ingold	Willoughby and Archer (1973); Sanders and Anderson (1979); Descals and Webster (1980)
Dendrospora fastuosa Descals and Webster	Descals and Webster (1981)
Dendrospora fusca Descals and Webster	Descals and Webster (1981)
Dendrospora tenella Descals and Webster	Descals and Webster (1981)
Dendrospora torulosa Descals and Webster	Descals and Webster (1981)
Dimorphospora foliicola Tubaki	Willoughby and Archer (1973); Révay and Gönczöl (1990); Shearer and Webster (1990)
Diplocladiella scalaroides Arnaud ex. M.B. Ellis	Shearer and von Bodman (1983); Révay and Gönczöl (1990)
Dwayaangam cornuta Descals	Descals and Webster (1982)
Filosporella annelidica (Shearer and Crane) Crane and Shearer	Shearer and von Bodman (1983); Révay and Gönczöl (1990); Shearer and Bartolata (unpublished)
Flabellospora acuminata Descals	Archer and Willoughby (1969) (as unnamed species of *Flabellospora*, Descals and Webster 1982)
Flabellospora amphibia (Price and Talbot) Descals	Price and Talbot (1966) (as *Tetracrium amphibium* Price and Talbot); Descals and Webster (1982)
Flagellospora curvula Ingold	Willoughby and Archer (1973); Sanders and Anderson (1979); Shearer and Webster (1991)
Flagellospora minuta Iqbal and Bhatty	Iqbal and Bhatty (1980)
Flagellospora penicillioides Ingold	Sanders and Anderson (1979)
Fontanospora eccentrica (Petersen) Dyko	Khattab and Webster (unpublished) – Oak and alder twigs, River Teign, Devon, England
Geniculospora inflata (Ingold) Nils. ex Marv. and Nils.	Webster and Descals (1981)
Goniopila monticola (Dyko) Marv. and Descals	Shearer and Webster (1991)
Heliscella stellata (Ingold and Cox) Marv.	Willoughby and Archer (1973) (as *Clavatospora stellata*); Révay and Gönczöl (1990)
Heliscella stellatacula Kirk ex Marv. and Nils.	Kirk (1969); Jones (1972)

Table 4.2 (Continued)

Species	References
Heliscina campanulata Marv.	Shearer and Webster (1991)
Heliscus lugdunensis Sacc. and Therry	Sládečková (1963); Eaton and Jones (1971); Willoughby and Archer (1973); Lamore and Goos (1978); Kane (1980); Révay and Gönczöl (1990); Shearer and Webster (1991)
Isthmotricladia britannica Descals	Descals and Webster (1982); Révay and Gönczöl (1990)
Laridospora appendiculata (Anastasiou) Nawawi	Nawawi 1976
Lemonniera alabamensis Sinclair and Morgan-Jones	Sinclair and Morgan-Jones (1979)
Lemonniera aquatica de Wildeman	Sládečková (1963); Jones and Oliver (1964); Jones (1972); Willoughby and Archer (1973); Sanders and Anderson (1979); Révay and Gönczöl (1990); Shearer and Bartolata (unpublished)
Lemonniera cornuta Ranzoni	Ranzoni (1953)
Lemonniera terrestris Tubaki	Willoughby and Archer (1973); Sanders and Anderson (1979); Révay and Gönczöl (1990)
Lunulospora curvula Ingold	Sládečková (1963); Shearer and Webster (1991); Shearer and Bartolata (unpublished)
Mycocentrospora acerina (Hartig) Deighton	Willoughby and Archer (1973); Shearer and Webster (1991)
Mycocentrospora angulata (Petersen) Iqbal	Sanders and Anderson (1979)
Mycocentrospora aquatica (Iqbal) Iqbal	Iqbal (1971)
Mycocentrospora clavata Iqbal	Iqbal (1974)
Quadricladium aquaticum Nawawi and Kuthubutheen	Nawawi and Kuthubutheen (1989)
Pyramidospora casuarinae S. Nilss.	Nilsson (1962)
Sigmoidea aurantiaca Descals	Descals and Webster (1982); Révay and Gönczöl (1990)
Taeniospora gracilis Marv.	Willoughby and Archer (1973); Sanders and Anderson (1979)
Tetrachaetum elegans Ingold	Jones and Oliver (1964); Willoughby and Archer (1973); Sanders and Anderson (1979)
Tetracladium marchalianum de Wildeman	Jones and Oliver (1964); Willoughby and Archer (1973); Sanders and Anderson (1979); Shearer and von Bodman (1983); Révay and Gönczöl (1990); Shearer and Bartolata (unpublished)
Tetracladium maxilliforme (Rostrup) Ingold	Webster and Descals (1981)
Tetracladium setigerum (Grove) Ingold	Sládečková (1963); Willoughby and Archer (1973); Révay and Gönczöl (1990)

Table 4.2 (Continued)

Species	References
Tricladium angulatum Ingold	Shearer and von Bodman (1983); Révay and Gönczöl (1990); Shearer and Bartolata (unpublished)
Tricladium biappendiculata (Arnold) Marv. and Descals	Arnold (1970)[b]
Tricladium chaetocladium Ingold	Shearer and Webster (1991)
Tricladium castaneicola Sutton	Webster and Descals (1981)
Tricladium chaetocladium Ingold	Shearer and Webster (1991)
Tricladium giganteum Iqbal	Sanders and Anderson (1979); Shearer and Webster (1991)
Tricladium gracile Ingold	Sanders and Anderson (1979); Shearer and Webster (1991)
Tricladium splendens Ingold	Jones and Oliver (1964); Eaton and Jones (1971); Jones (1972); Willoughby and Archer (1973); Sanders and Anderson (1979); Révay and Gönczöl (1990); Shearer and Webster (1991)
Tricladium terrestre Park	Webster (unpublished) – Branch of *Ulex* submerged in the River Swincombe, Devon, England
Tricladium varium Jones	Jones and Stewart (1972)
Tricladium state of *Hymenoscyphus varicosporoides* Tubaki	Tubaki (1966)
Tripospermum camelopardus	Willoughby and Archer (1973)
Triposporina ceranoica sp. ined. Descals	Webster and Descals (1981)
Triscelophorus acuminatus Nawawi	Nawawi (1975)
Triscelophorus monosporus Ingold	Shearer and von Bodman (1983); Shearer and Webster (1991)
Tumularia aquatica (Ingold) Descals and Marv.	Sládečková (1963) (as *Dactylella aquatica*); Willoughby and Archer (1973) (as *D. aquatica*); Kane (1980) (as *D. aquatica*); Révay and Gönczöl (1990); Shearer and Webster (1991)
Tumularia tuberculata (Gönczöl) Descals and Marv.	Révay and Gönczöl (1990)
Vargamyces aquaticus (Dudka) Toth	Révay and Gönczöl (1990); Shearer (unpublished) – Twigs submerged in Jordan Creek, Vermilion County, Illinois
Varicosporium elodeae Kegel	Willoughby and Archer (1973); Sanders and Anderson (1979)
Volucrispora graminea (Ingold, McDougall and Dann)	Sanders and Anderson (1979)

[a] All collections from Shearer and Bartolata (unpublished) are from birchwood baits placed in Jordan Creek, Vermilion County, Illinois.
[b] From wood in terrestrial habitats.

Table 4.3. Aquatic hyphomycetes listed as predominant species on submerged wood in studies in which species were quantified by frequency of occurrence or conidial production

	Willoughby and Archer (1973)	Sanders and Anderson (1979)	Shearer and Webster (1991)	Révay and Gönczöl (1990)	Shearer and Bartolata (unpublished)
Heliscus lugdunensis[a]	x		x		
Anguillospora longissima[a]	x		x	x	x
Clavariopsis aquatica[a]	x		x	x	x
Lemonniera aquatica	x				
Alatospora acuminata	x		x	x	x
Dimorphospora foliicola[a]	x		x	x	
Tricladium splendens[a]	x	x	x		
Flagellospora curvula		x	x		
Tricladium gracile		x			
Dendrospora erecta		x			
Casaresia sphagnorum		x			
Articulospora tetracladia[a]		x			
Tricladium giganteum		x			
Varicosporium elodeae		x			
Mycocentrospora acerina			x		
Tricladium chaetocladium[a]			x		
Tumularia aquatica[a]			x		
Anguillospora crassa[a]			x		
Lunulospora curvula			x		x
Anguillospora furtiva[a]				x	
Vargamyces aquaticus				x	
Filosporella annelidica[a]				x	x
Tetracladium marchalianum					x

[a] Species connected to a teleomorphic state.

gradually disappear with increasing time of submergence. These species may have ruderal strategies in that they colonize and sporulate quickly, and then disappear or decline in importance over time.

4.3 Role of Woody Debris in Aquatic Hyphomycete Life Cycles

4.3.1 Habitat and Nutritional Resource

Clearly a large number of aquatic hyphomycetes use wood as a habitat (see Table 4.2). There is also evidence that some of these species are able to use wood as a nutritional resource. Wood is a complex of many types of chemicals whose concentrations vary among and within species, within parts of a single tree (e.g. sapwood, heartwood, and bark), and with height (Browning 1963; Findlay 1985). Chemical constituents of wood can be grouped into (1) primary metabolites such as soluble sugars, lipids, peptides and the major storage compound, starch; (2) structural cell wall components such as hemicellulose, cellulose and lignin; (3) minerals; and (4) extractives (exclusive of extractable primary metabolites). The chemistry and occurrence of these constituents and the anatomical structure of wood as they relate to decomposition have been discussed in detail elsewhere (Kirk 1983; Kirk and Cowling 1984; Pettersen 1984; Harmon et al. 1986; Rayner and Boddy 1988) and will not be considered here. Rather the general importance of these classes of constituents to aquatic hyphomycetes and the ability of aquatic hyphomycetes to use these constituents will be discussed.

Seventeen published reports of enzyme activity and/or wood decomposition ability by some of the aquatic hyphomycetes listed as important on wood in Table 4.3 exist (Table 4.4). Generally most of these species are able to use primary metabolites, particularly, starch. Primary metabolites, however, comprise a small percentage of wood and are usually limited to the sapwood parenchyma of living or recently dead wood (Rayner and Boddy 1988). It is likely that primary metabolites are easily leached from wood in water and readily assimilated by heterotrophs. Thus these nutrients may be important only for fungi that colonize newly fallen wood that enters water directly.

Of the three major structural materials, hemicelluloses are the most easily hydrolyzed. They comprise 25–40% of temperate angiosperm woods and 25–30% of coniferous woods (Rayner and Boddy 1988). The major hemicellulose of angiosperms is glucuronoxylan (15–30%) followed by glucomannan (2–5%) (Pettersen 1984). In gymnosperms, glucomannans predominate (15–23%) followed by arabinoglucuronoxylan (7–10%). The ability to degrade xylan has been tested for numerous lignicolous species, but surprisingly, not all of these species were able to hydrolyze it (Table 4.4). Three species unable to degrade xylan were, however, able to use xylose. Less is known about the ability of species to hydrolyze glucomannans. Most likely this lack of knowledge results from the lack of com-

Table 4.4. The enzymatic capabilities of some aquatic hyphomycetes that occur frequently and/or abundantly on wood

Species and pertinent references	Crystalline cellulose	Substrate							Wood (weight loss)	Wood (soft-rot cavities)
		CMC	Cellobiose	Amylose	Xylan	Xylose	Lignin	Pectin		
Heliscus lugdunensis [4,6,7,10,11,13,15,16,17][a]	+	(+,−)	+	+	(+,−)	(+,−)	(+,−)	+	+	(+,−)
Anguillospora longissima [4,6,9,17]	(+,−)	+	+	+	+	+	(+,−)	(+,−)	+	+
Clavariopsis aquatica [4,5,10,11,12,15,17]	+	+	+	+	+	+	(+,−)	+ (PG,PL)	+	+
Lemonniera aquatica [2,3,5,10,11,12,17]	(+,−)	+	+	+	+	+	+	+ (PG,PL)	ND	ND
Alatospora acuminata [9,11,12,17]	+	+	+	+	+	ND	+	+ (PG,PL)	ND	ND
Dimorphospora foliicola [4,14]	−	ND	ND	(+,−)	ND	ND	+	+	ND	ND
Tricladium splendens [1,2,3,5,6,13,17]	+	(+,−)	+	+	+	+	+	+ (PG,PL)	+	+
Flagellospora curvula [10,11,12]	+ (weak)	+	ND	ND	+	ND	ND	+ (PG,PL)	ND	ND
Tricladium gracile [4]	ND	ND	ND	ND	ND	ND	−	ND	ND	ND
Dendrospora erecta [17]	ND	+	+	+	+	+	+	+	ND	ND
Articulospora tetracladia [1,2,4,6,13,17]	−	+	+	+	+	+	(+,−)	+ (PG,PL)	+	−
Tricladium giganteum [4]	ND	ND	ND	ND	ND	ND	+	ND	ND	ND
Varicosporium elodeae [1,2,13,14,17]	−	(+,−)	(+,−)	+	−	+	+	+ (PG,PL)	ND	ND

Tricladium chaetocladium [1]	+	+	+	ND	ND	ND	ND	ND	ND	ND
Tumularia aquatica [4,5,17]	–	–	+	+	–	+	+	+	+	ND
Anguillospora crassa [4,6,17]	ND	+	+	+	–	+	+	+	+	+
Lunulospora curvula [4,6,12]	+	ND	ND	ND	ND	ND	–	ND	+	+
Anguillospora furtiva [4]	ND	ND	ND	ND	ND	ND	+	ND	ND	ND
Filosporella annelidica [10,15,16]	+	+	+	+	+	ND	–	+ (PG,PL)	+	(+,–)
Tetracladium marchalianum [1,3,8,11,12,13]	(+,–)	+	+	+	(+,–)	+	+	+ (PG,PL)	ND	ND

( ) = conflicting data; ND = no data available; CMC = carboxymethyl cellulose; BD = lignosulfonates with wood sugars; PG = polygalacturonase; PL = pectin lyase.

[a] *References*: [1] Chamier (1985); [2] Chamier and Dixon (1982); [3] Danninger et al. (1979); [4] Fisher et al. (1983); [5] Gunasekera and Webster (1983); [6] Jones (1981); [7] Leightley and Eaton (1977); [8] Nilsson (1964); [9] Ranzoni (1951); [10] Suberkropp et al. (1983); [11] Suberkropp and Klug (1980); [12] Suberkropp and Klug (1981); [13] Thornton (1963); [14] Tubaki (1958); [15] Zare-Maivan and Shearer (1988a); [16] Zare-Maivan and Shearer (1988b); [17] Zemek et al. (1985).

mercially available glucomannan; the compound must be extracted from wood and purified. Only one species, *H. lugdunensis*, has been grown on glucomannan and this species caused clearing of the glucomannan medium (Leightley and Eaton 1977). Ability to degrade hemicelluloses in leaves was demonstrated for *Tetracladium marchalianum* and *Flagellospora curvula* (Suberkropp and Klug 1980), but not for *Lemonniera aquatica*, *Clavariopsis aquatica* and *Alatospora acuminata*. Underestimation of mannan degrading ability may have occurred due to the presence of fungal mannans in the test fungi.

The most abundant polymer in wood, cellulose, may account for about 40–50% of the dry weight of temperate woods (Pettersen 1984). Native cellulose is partially crystalline and requires three hydrolytic enzymes acting synergistically for its complete degradation (Kirk and Cowling 1984). Endogluconase and exogluconase break down cellulose into oligosaccharides and the disaccharide cellobiose which is converted to glucose by cellobiose. Endoglucanase acts only on non-crystalline cellulose such as carboxymethyl cellulose while exogluconase acts on crystalline and non-crystalline cellulose. All of the lignicolous aquatic hyphomycetes that have been tested were able to degrade cellobiose while all but one were able to degrade non-crystalline cellulose. Eight species were able to degrade crystalline cellulose and there were conflicting data from different studies for three species. Species unable to degrade crystalline cellulose may rely on other species to condition native cellulose before they can use it or must use some other wood component as a carbon source.

Lignins are complex aromatic polymers formed from phenyl propanoid subunits (Kirk and Obst 1988). They comprise 20–30% of woody tissue (Pettersen 1984) and form a physical barrier that protects cellulose and hemicellulose from degradative enzymes. Cellulose microfibrils are coated or overlayered by hemicelluloses which are covered by a lignin sheath (Kirk and Cowling 1984). The complex structure of lignin makes it difficult for microbes to degrade, and decay rates of submerged wood are inversely related to lignin content (Melillo et al. 1983). Lignicolous aquatic hyphomycetes have been tested for their ability to degrade extracted lignins (Table 4.4). Nine of 17 species tested were positive and there are conflicting data for three species. It is very difficult to assess lignolytic activity with agar plate techniques (Fisher et al. 1983) and the extracted lignins which were used in past studies (Fisher et al. 1983, Zare-Maivan and Shearer 1988a) are highly modified forms of native lignin (Kirk and Obst 1988). Because of the limitations of these techniques, the lignin-degrading capabilities of aquatic hyphomycetes are not known.

The negative influence of wood extractives on terrestrial decomposer species is well known (Scheffer and Cowling 1966). Little is known, however, regarding the effects of wood extractives on aquatic hyphomycetes. Bärlocher and Oertli (1978) found that extracts of conifer needles were able to depress the growth of five aquatic hyphomycete species and Bärlocher et al. (1978) demonstrated that extracted pine needles were colonized more readily by aquatic hyphomycetes than non-extracted needles. Toxic extractives are known to be leached from wood in terrestrial habitats (Scheffer and Cowling 1966), and it is likely that this process is accelerated in aquatic habitats. Rapid leaching, coupled with gradual deactivation of toxic compounds due to antigenous enzymes and microbial activity, suggest that extractives have their greatest effect in the initial colonization of fresh-fallen wood.

Apparently, most of the lignicolous aquatic hyphomycetes tested thus far have the enzyme capabilities to degrade wood (see Table 4.4). For some species such as *Heliscus lugdunensis*, however, the production of enzymes could not be quantitatively correlated with activity on natural substrata (Suberkropp et al. 1983; Butler and Suberkropp 1986; Zare-Maivan and Shearer 1988b). Jones (1981) found that four of six lignicolous hyphomycetes were able to cause weight-loss and soft-rot cavities on wood test blocks. *Articulospora tetracladia* and *H. lugdunensis* caused little weight-loss and formed no soft-rot cavities. In a recent study of the wood-rotting ability of 22 aquatic hyphomycetes (Shearer unpublished), species fell into three distinct groups. The first group includes species that caused extensive weight-loss and formed soft-rot cavities. The second group caused moderate weight-loss and formed no soft-rot cavities while the third group caused very little weight loss and formed no soft-rot cavities. Most of the species in the first group (e.g. *Anguillospora crassa*, *A. longissima*, *Clavariopsis aquatica*, and *Lunulospora curvula*) have been reported frequently from submerged wood. Two species that have not been reported from wood (*Clavatospora longibrachiata* and *M. aquatica*) occurred in the third group along with two species that have been reported from wood (*Filosporella annelidica* and *Varicosporium elodeae*). It is likely that the latter two species may rely on dissolved carbon sources or products from the enzymatic activities of other fungi. Species in the second group, such as *H. lugdunensis*, may degrade parenchyma cells, pectins and wood components other than secondary walls. There is evidence that *H. lugdunensis* is an early successional species (Willoughby and Archer 1973; Shearer and Webster 1991) and its inability to extensively degrade secondary walls may cause its disappearance when other substrates are exhausted.

4.3.2 Site of Sexual Reproduction

Sexual states have not been found for the majority of aquatic hyphomycetes (Chap. 5). Those species with sexual states may be either Ascomycetes or Basidiomycetes. A number of these sexual states have been reported from wood (Ranzoni 1956; Webster 1959; Willoughby and Archer 1973; Webster and Descals 1981; Abdullah et al. 1981; Hamad and Webster 1987). Interestingly, of the 23 anamorph species reported as quantitatively important on wood, 11 have been connected to sexual states (Table 4.3). Since sexual states have been reported for only 20 out of about 300 species of aquatic hyphomycetes, such species are disproportionately represented on wood. It is interesting to speculate on whether this phenomenon represents a true evolutionary difference among species or is an effect of the substratum or its handling. The aquatic hyphomycetes are clearly a polyphyletic group with evolutionary links to the Basidiomycetes, Discomycetes, Pyrenomycetes and Loculoascomycetes. It would not be unreasonable to expect one line to have developed from terrestrial lignicolous Discomycetes, another from terrestrial Basidiomycetes, a third from terrestrial foliicolous Deuteromycetes, and so on. Parallel evolution in conidial modification to aquatic habitats may have occurred in unrelated taxa on a variety of substrata. Lignicolous aquatic hyphomycetes may have been pre-adapted to the lignicolous substrate and this may explain why they are frequent and successful wood colonizers.

The possibility also exists that some feature of wood not characteristic of leaves, e.g. its persistence or nutrient content, permits or stimulates the development of sexual states. Experimental evidence to support this idea is at present lacking. Before we place too much emphasis on wood as the site of sexual reproduction for aquatic hyphomycetes, we also need to examine whether this phenomenon is a sampling artifact. Traditionally, wood has been incubated in moist chambers while leaves have been incubated in water. Some sexual states may not be formed under water and therefore go undetected on submerged leaves. The question arises, if we incubated leaves in moist chambers, would we find more sexual states on them? Evidence suggests this might be so since two sexual states, *Hymenoscyphus tetracladius* Abdullah, Descals and Webster (Abdullah et al. 1981), and *Hymenoscyphus africanus* Descals, Fisher and Webster (Descals et al. 1984) were described from leaves incubated in moist chambers.

Whatever the evolutionary history, sexual states are rarely found on submerged leaves in nature (one exception, however, is *Aquadiscula aquatica* Shearer and Crane, which has been collected only from leaves and appears to lack an aquatic hyphomycete anamorph; Shearer and Crane 1985). Thus wood may play an important role as a site for genetic recombination in aquatic hyphomycetes. In addition, the discomycetous teleomorphs are adapted for terrestrial spore discharge and dispersal. When submerged woody debris becomes partially exposed to air, spores may enter air currents and be dispersed over wide areas. Thus the ability of an aquatic hyphomycete to form a sexual fruiting body specialized for wind dispersal of spores may contribute to the long distance dispersal of that fungus. Interestingly, Willoughby and Archer (1973) found that apothecia of *Cudoniella clavus* (Albertini and Schweinitz:Fr.) Dennis on submerged twigs were brought to the surface of water by stipes of various lengths. The stipes of fruit bodies on wood incubated in moist chambers were always short. It should be noted that submerged wood is also the habitat for numerous teleomorphs, mostly Ascomycetes, that lack aquatic hyphomycete anamorphs (Shearer and Crane 1978, Shearer and von Bodman 1983, Shearer 1989a,b; Chap. 5).

4.3.3 Site of Interspecific and Intraspecific Interactions

Much is known about the interactions between grazing invertebrates and aquatic hyphomycetes on submerged leaves (Bärlocher 1985; Chap. 6). Newly fallen leaves are rapidly colonized by aquatic hyphomycetes, which increase the nutritional quality and palatability of most leaf species. Colonized leaf material is then grazed on by stream invertebrates. Little is known about a comparable system on wood. Woody debris supports a diverse invertebrate faunal assemblage (Nilsen and Larimore 1973; Anderson et al. 1978; Dudley and Anderson 1982). Dudley and Anderson (1982) have separated wood-associated invertebrates into five functional groups based on how they use wood: 1: boring or tunnelling; 2: ingestion of wood by gouging, scraping or rasping; 3: scraping of associated aufwuchs; 4: use as a physical habitat; 5: predation. Only invertebrates in groups 1 and 2 actually ingest wood. Dudley and Anderson (1982) indicate that microbial conditioning and the resultant softening of the wood surface are important because this softening permits abrasion and ingestion by scrapers. At the softened stage, the wood also becomes more suitable for obligate wood gougers and more generalized shredders.

Dudley and Anderson (1982) further suggest that borers are partially dependent on microbes for nutrition and that as they bore through wood, they open it up for microbial colonization. It is likely that aquatic hyphomycetes with soft rot capabilities are very important in the initial conditioning of submerged wood. Colonization of submerged wood by such aquatic hyphomycetes occurs within three weeks of submersion (Shearer and Bartolata 1990). Pereira et al. (1982) have found fungal hyphae and spores associated with wood in the gut contents of wood borers, gougers and grazers. Research is needed to determine how aquatic hyphomycetes alter the nutritional quality, palatability and physical texture of wood and how invertebrates influence the fungal colonization of wood and fungal community structure. In a preliminary study, the invertebrate grazer, *Asellus*, was maintained successfully on birchwood sticks colonized by various species of aquatic hyphomycetes and ascomycetes until experiment termination after several months (Shearer and Mattingly, unpubl. obs.). Animals on uncolonized sticks (control) died within three to four weeks. During the course of the study, offspring were produced and fed successfully on the fungi. Fungal biomass was continually cropped at the wood surface, but there was no evidence of wood gouging.

It is thought that substrata such as woody debris which are stable over time and decompose slowly favour colonization by competitive (Pugh 1980) or combative (Cooke and Rayner 1984) fungi. Such species have the ability to defend captured resources and/or to invade already colonized substrata by the production of antimicrobial agents or direct hyphal interference (Wicklow 1981). Shearer and Zare-Maivan (1988) tested the hypothesis that lignicolous aquatic Ascomycetes and Fungi Imperfecti were more antagonistic than foliicolous species. They found that lignicolous species were more inhibitory to, and resistant to inhibition by, other species than foliicolous species. Several aquatic hyphomycetes common on wood, e.g. *Clavariopsis aquatica*, *Anavirga dendromorpha*, and *Tetracladium marchalianum* were among the most competitive species. *Filosporella annelidica*, a species reported from wood and leaves, however, was the poorest competitor among 25 species.

Competitive ability in culture is not always a good prognosticator of competitive ability in the field (Webber and Hedger 1986). Shearer and Bartolata (1990) tested the ability of some antagonistic and non-antagonistic lignicolous fungi to defend colonized wood in situ by pre-colonizing birchwood sticks with test species and then submerging the sticks in a stream. An unidentified aquatic hyphomycete with sigmoid spores (S-77-9D), which was isolated from submerged wood, was able to exclude five out of six potential invaders for 4 weeks and four of seven potential invaders for 8 weeks when the experiment was terminated. In addition to preventing colonization, this species also decreased sporulation of invading species compared to controls. Parallel studies using autoclaved and non-autoclaved sticks pre-colonized by this species revealed that the presence of the living fungus, not just prior colonization by the fungus, was required for exclusion of invaders. In contrast, *F. annelidica*, a poor competitor in culture, was unable to exclude most potential invaders. Several lignicolous ascomycetes found to be antagonistic in culture (Shearer and Zare-Maivan 1988) were also able to exclude aquatic hyphomycetes in situ (Shearer and Bartolata 1990).

Intraspecific fungal incompatibility is a conspicuous feature of terrestrial lignicolous Basidiomycetes (Rayner and Todd 1979). Barriers between incompatible strains of the same fungus can be seen in culture and in decayed wood. A study of intraspecific interactions among aquatic hyphomycetes (Bärlocher 1991) suggests that the aggressive incompatibility characteristic of lignicolous basidiomycetes may be absent in lignicolous aquatic hyphomycetes.

4.3.4 Reservoir

The input of leaves, a major carbon source for aquatic hyphomycetes, is mostly restricted to the autumn. Because leaves of most tree species are decomposed within one year, the quantity of leaf litter decreases dramatically in the spring and early summer. In contrast, the input of woody debris is more episodic than seasonal and enters water as a result of wind-throw,

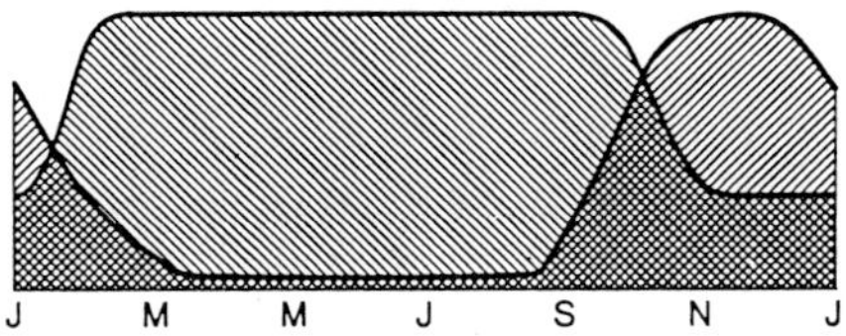

Fig. 4.1. Relative importance of leaves and wood throughout a year

storm activity or ice damage. Woody debris of large size resists displacement by water flow, even during flood conditions and is decomposed very slowly. These characteristics make wood a more temporally constant substratum for aquatic hyphomycetes than leaves. This means that the relative importance of woody debris to aquatic hyphomycetes increases in the early spring and continues through the summer until the autumnal leaf-fall (Fig. 4.1). At this time wood may serve as a source of inoculum of aquatic hyphomycetes for newly fallen leaves. It may have the same function after spates which have scoured leaves from the stream bed.

4.3.5 Long Distance Dispersal

One of the most intriguing aspects of the aquatic hyphomycetes is their world-wide distribution (Webster and Descals 1981). Woody debris may be an important agent for long distance dispersal. Because of its persistence, wood can be transported long distances by successive floods, eventually to be discharged into oceans. It is not unreasonable to expect that fungal hyphae associated with the wood would also be transported. Experimental evidence is needed to determine whether aquatic hyphomycetes are able to survive passage through environmentally unfavourable areas to grow and reproduce again when conditions are once again favourable. Mention has been made previously (Sect. 4.1.3.3) of the long distance wind dispersal of sexual propagules produced on wood.

4.4 Conclusions

For aquatic hyphomycetes, woody debris serves as a habitat, reservoir, nutritional resource, site of asexual and sexual reproduction and competition, and vehicle for long-distance dispersal. The influence of wood characteristics on aquatic hyphomycete community dynamics are varied (see Table 4.1B). Woody debris tends to be a more selective substratum than leaves and thus initial colonization patterns may be more deterministic than stochastic. Because of the persistence of the substratum, it lasts through different seasonal environmental conditions and different communities develop over time. Hence colonization patterns of individual aquatic hyphomycete species are more successive than parallel. Taxonomic diversity of fungi colonizing woody debris is broad and includes flagellated fungi, dematiaceous hyphomycetes, a variety of Ascomycetes and several Basidiomycetes in addition to the aquatic hyphomycetes. This diversity may be related to substratum persistence. Also related to substratum persistence may be the high degree of competitiveness found for lignicolous fungi and the high teleomorph:anamorph ratio. The low rate of decomposition accounting, in part, for the substratum persistence is most likely related to the low surface to volume ratio and high density which limits oxygen diffusion into water-soaked substrates. Lack of oxygen within the substratum restricts aerobic heterotroph activities to the surface of woody debris. Also the high concentrations of polymeric carbon sources and the presence of phenolics make wood a difficult substrate to digest for invertebrates and microbes.

Clearly, more research on all aspects of the interaction between aquatic hyphomycetes and wood is needed. At the same time, the fungal-wood system makes a very interesting model with which to study fungal ecology.

References

Abdullah SK, Descals E, Webster J (1981) Teleomorphs of three aquatic hyphomycetes. Trans Br Mycol Soc 77:475–483

Anderson NH, Sedell JR (1979) Detritus processing by macroinvertebrates in stream ecosystems. Annu Rev Entomol 24:351–377

Anderson NH, Sedell JR, Roberts LM, Triska FJ (1978) The role of aquatic invertebrates in processing of wood debris in coniferous forest streams. Am Midl Nat 100:64–82

Archer J, Willoughby LG (1969) Wood as a growth substratum for a freshwater foam spore. Trans Br Mycol Soc 53:484–487

Arnold GRW (1970) De Specie nova generis *Ingoldia*. Nov Syst Plant non Vasc 6:173–176

Aumen NG, Bottomly PJ, Ward GM, Gregory SV (1983) Microbial decomposition of wood in streams: distribution of microflora and factors affecting [^{14}C]lignocellulose mineralization. Appl Environ Microbiol 46:1409–1416

Bärlocher F (1985) The role of fungi in the nutrition of stream invertebrates. Bot J Linn Soc 91:83–94

Bärlocher F (1991) Intraspecific hyphal interactions among aquatic hyphomycetes. Mycologia 83:82–88

Bärlocher F, Oertli JJ (1978) Inhibitors of aquatic hyphomycetes in dead conifer needles. Mycologia 70:964–974

Bärlocher F, Schweizer M (1983) Effects of leaf size and decay rate on colonization by aquatic hyphomycetes. Oikos 41:205–210

Bärlocher F, Kendrick B, Michaelides J (1978) Colonization and conditioning of *Pinus resinosa* needles by aquatic hyphomycetes. Arch Hydrobiol 81:462–474

Benner R, Moran MA, Hodson RE (1986) Biogeochemical cycling of lignocellulosic carbon in marine and freshwater ecosystems: relative contributions of prokaryotes and eukaryotes. Limnol Oceanogr 31:89–100

Bilby RE, Likens GE (1980) Importance of organic debris dams in the structure and function of stream ecosystems. Ecology 61:1107–1113

Bird GA, Kaushik NK (1981) Coarse particulate organic matter in streams. In: Lock MA, Williams DD (eds) Perspectives in running water ecology. Plenum, New York, pp 41–68

Browning BL (1963) The composition and chemical reactions of wood. In: Browning BL (ed) The chemistry of wood. Interscience Publ, John Wiley and Sons, New York, pp 57–102

Butler SK, Suberkropp K (1986) Aquatic hyphomycetes on oak leaves: comparison of growth, degradation and palatability. Mycologia 78:922–928

Chamier AC (1985) Cell-wall degrading enzymes of aquatic hyphomycetes: a review. Bot J Linn Soc 91:67–81

Chamier AC, Dixon PA (1982) Pectinases in leaf degradation by aquatic hyphomycetes: the enzymes and leaf maceration. J Gen Microbiol 128:2469–2483

Cooke RC, Rayner ADM (1984) Ecology of saprotrophic fungi. Longman, New York

Crane JL, Dumont KP (1975) Hyphomycetes from the West Indies and Venezuela. Can J Bot 53:843–851

Danninger E, Messner K, Rohr M (1979) Untersuchungen über den biologischen Abbau organischer Naturstoffe durch aquatische Hyphomyzeten. Zentralbl Bakteriol Hyg I Abt Orig B 169:282

Descals E, Sutton BC (1976) *Anavirga dendromorpha* and its *Phialocephala* phialidic state. Trans Br Mycol Soc 67:269–274

Descals E, Webster J (1982) Taxonomic studies on aquatic hyphomycetes III. Some new species and a new combination. Trans Br Mycol Soc 78:405–437

Descals E, Fisher PJ, Webster J (1984) The *Hymenoscyphus* teleomorph of *Geniculospora grandis*. Trans Br Mycol Soc 83:541–546

Dudley T, Anderson NH (1982) A survey of invertebrates associated with wood debris in aquatic habitats. Melanderia 39:1–21

Eaton RA, Jones EBG (1971) The biodeterioration of timber in water cooling towers. II. Fungi growing on wood in different positions in water cooling towers. Mater Org 6:81–92

Findlay WPK (ed) (1985) Preservation of timber in the tropics. Dr. W Junk Publishers, Dordrecht, The Netherlands

Fisher PJ, Davey RA, Webster J (1983) Degradation of lignin by aquatic and aero-aquatic hyphomycetes. Trans Br Mycol Soc 80:166–168

Fisher SG, Likens GE (1972) Stream ecosystem: organic energy budget. BioScience 22:33–35

Gunasekera SA, Webster J (1983) Effect of nitrate and phosphate on weight losses of pine and oak wood caused by aquatic and aero-aquatic hyphomycetes. Trans Br Mycol Soc 80:507–514

Hamad SR, Webster J (1987) *Anavirga dendromorpha*, anamorph of *Apostemidium torrenticola*. Sydowia 40:60–64

Harmon ME, Franklin JF, Swanson FJ, Sollins P, Gregory SV, Lattin JD, Anderson NH, Cline SP, Aumen NG, Sedell JR, Lienkaemper GW, Cromack K JR, Cummins KW (1986) Ecology of coarse woody debris in temperate ecosystems. Rec Adv Ecol Res 15:133–302

Ingold CT (1952) *Actinospora megalospora* n. sp., an aquatic hyphomycete. Trans Br Mycol Soc 35:66–70

Iqbal SH (1971) New aquatic hyphomycetes. Trans Br Mycol Soc 56:343–352

Iqbal SH (1974) New aquatic hyphomycetes. Biologia 20:1–10

Iqbal SH, Bhatty SF (1980) New freshwater hyphomycetes from Pakistan. Trans Mycol Soc Jpn 21:71–75

Jones EBG (1972) The decay of timber in aquatic environments. Br Wood Preserv Assoc Annu Conv, pp 1–18

Jones EBG (1981) Observation on the ecology of lignicolous aquatic hyphomycetes. In: Wicklow DT, Carroll GC (eds) The fungal community: its organization and role in the ecosystem. Marcel Dekker, New York, pp 731–742

Jones EBG, Oliver AC (1964) Occurrence of aquatic hyphomycetes on wood submerged in fresh and brackish water. Trans Br Mycol Soc 47:45–48

Jones EBG, Stewart RJ (1972) *Tricladium varium*, an aquatic hyphomycete on wood in water-cooling towers. Trans Br Mycol Soc 59:163–167

Kane D (1980) The effect of sewage effluent on the growth of micro-organisms in the marine environment. PhD Thesis, Portsmouth Polytechnic, Portsmouth, England

Kirk PW (1969) Aquatic hyphomycetes on wood in an estuary. Mycologia 61:177–181

Kirk TK (1983) Degradation and conversion of lignocellulose. In: Smith JF, Berry DR, Kristansen B (eds) The filamentous fungi. Edward Arnold, London, pp 266–295

Kirk TK, Cowling EB (1984) Biological decomposition of solid wood. Am Chem Soc Ser 207:457–487

Kirk TK, Obst JR (1988) Lignin determination. In: Wood WA, Kellogg ST (eds) Methods in enzymology – biomass, part b. Lignin, pectin, and chitin. Academic Press, San Diego, 161:87–101

Lamore BJ, Goos RD (1978) Wood-inhabiting fungi of a freshwater stream in Rhode Island. Mycologia 70:1025–1034

Leightley LE, Eaton RA (1977) Mechanisms of decay of timber by aquatic micro-organisms. Br Wood Preserv Assoc Annu Conv, pp 1–16

Melillo JM, Naiman RJ, Aber JD, Eshleman KN (1983) The influence of substrate quality and stream size on wood decomposition dynamics. Oecologia 58:281–285

Nawawi A (1974) A new *Campylospora*. Trans Br Mycol Soc 63:603–607

Nawawi A (1975) *Triscelophorus acuminatus* sp. nov. Trans Br Mycol Soc 64:345–348

Nawawi A (1976) A new genus of Hyphomycetes. Trans Br Mycol Soc 66:344–347

Nawawi A, Kuthubutheen AJ (1988) Additions to *Condylospora* (Hyphomycetes) from Malaysia. Mycotaxon 33:329–338

Nawawi A, Kuthubutheen AJ (1989) *Quadricladium aquaticum* Gen et sp. nov., an aquatic hyphomycete with tetraradiate conidia. Mycotaxon 34:489–495

Nilsen HC, Larimore RW (1973) Establishment of invertebrate communities on log substrates in the Kaskaskia River, Illinois. Ecology 54:366–374

Nilsson S (1964) Freshwater hyphomycetes: taxonomy, morphology and ecology. Symb Bot Ups 18:1–130

Pereira CRD, Anderson NH, Dudley T (1982) Gut content analysis of aquatic insects from wood substrates. Melanderia 39:23–33

Perrott E (1960) *Ankistrocladium fuscum* Gen nov., sp. nov., an aquatic hyphomycete. Trans Br Mycol Soc 43:556–558

Petersen RH (1960) *Culicidospora*, a new genus of aquatic, aleuriosporous hyphomycetes. Bull Torrey Bot Club 87:342–347

Pettersen RC (1984) The chemical composition of wood. In: Rowell RM (ed) The chemistry of solid wood. Advances in chemistry series 207. American Chemical Society, Washington, DC, pp 57–126

Price IP, Talbot PHB (1966) An aquatic hyphomycete in a lignicolous habitat. Aust J Bot 14:19–23

Pugh GFJ (1980) Strategies in fungal ecology. Trans Br Mycol Soc 75:1–14

Ranzoni FV (1951) Nutrient requirements for two species of aquatic hyphomycetes. Mycologia 43:130–141

Ranzoni FV (1953) The aquatic hyphomycetes of California. Farlowia 4:353–398

Ranzoni FV (1956) The perfect stage of *Flagellospora penicillioides*. Am J Bot 43:13–17

Rayner ADM, Boddy L (1988) Fungal decomposition of wood. Its biology and ecology. John Wiley & Sons, New York

Rayner ADM, Todd NK (1979) Population and community structure and dynamics of fungi in decaying wood. Adv Bot Res 7:333–420

Révay A, Gönczöl J (1990) Longitudinal distribution and colonization patterns of wood-inhabiting fungi in a mountain stream in Hungary. Nova Hedwigia 51:505–520

Sanders PF, Anderson JM (1979) Colonization of wood blocks by aquatic hyphomycetes. Trans Br Mycol Soc 73:103–107

Scheffer TC, Cowling EB (1966) Natural resistance of wood to microbial deterioration. Annu Rev Phytopathol 4:147–170

Shearer CA (1972) Fungi of the Chesapeake Bay and its tributaries. III. The distribution of wood-inhabiting ascomycetes and fungi imperfecti of the Patuxent River. Am J Bot 59:961–969

Shearer CA (1989a) *Aniptodera* (Halosphaeriaceae) from wood in freshwater habitats. Mycologia 81:139–146

Shearer CA (1989b) *Pseudohalonectria* (Lasiosphaeriaceae), an antagonistic genus from wood in freshwater. Can J Bot 81:139–146

Shearer CA, Bartolata M (1990) Experimental determination of in situ competitive interactions among aquatic lignicolous fungi. In: Reisinger A, Bresinsky A (eds) Abstr 4th Int Mycol Congr, Regensburg, Germany. Botanical Institute, Univ Regensburg

Shearer CA, Crane JL (1978) The distribution of *Nais inornata*, a facultative marine ascomycete. Mycotaxon 7:443–452

Shearer CA, Crane JL (1985) *Aquadiscula appendiculata*, a new genus and species of Discomycetes from leaves submerged in a freshwater swamp. Mycologia 77:441–446

Shearer CA, Crane JL (1986) Illinois fungi. XII. Fungi and myxomycetes from wood and leaves submerged in southern Illinois swamps. Mycotaxon 25:527–538

Shearer CA, von Bodman S (1983) Patterns of occurrence of ascomycetes associated with decomposing twigs in a mid-western stream. Mycologia 75:518–530

Shearer CA, Webster J (1991) Aquatic hyphomycetes in the River Teign IV. Twig colonization. Mycol Res 95:413–420

Shearer CA, Zare-Maivan H (1988) In vitro hyphal interactions among wood- and leaf-inhabiting Ascomycetes and Fungi Imperfecti from freshwater habitats. Mycologia 80:31–37

Sinclair RC, Morgan-Jones G (1979) Notes on hyphomycetes. XXXII. Five new aquatic species. Mycotaxon 9:469–481

Sládečková A (1963) Aquatic deuteromycetes as indicators of starch campaign pollution. Int Rev Ges Hydrobiol 48:35–42

Suberkropp K, Klug MJ (1980) The maceration of deciduous leaf litter by aquatic hyphomycetes. Can J Bot 58:1025–1031

Suberkropp K, Klug MJ (1981) The degradation of leaf litter by aquatic hyphomycetes. In: Wicklow DT, Carroll GC (eds) The fungal community. Marcel Dekker, New York

Suberkropp K, Arsuffi TL, Anderson JP (1983) Comparison of degradative ability, enzymatic activity, and palatability of aquatic hyphomycetes grown on leaf litter. Appl Environ Microbiol 46:237–244

Swanson FJ, Lienkeamper GW, Sedell JR (1976) History, physical effects and management implications of large organic debris in western Oregon streams. USDA For Serv Gen Tech Rep PNW-56, Washington, DC

Swift MJ (1977) The ecology of wood decomposition. Sci Prog (Oxford) 64:175–199

Thornton DR (1963) The physiology and nutrition of some aquatic hyphomycetes. J Gen Microbiol 33:23–31

Triska FJ, Cromack K (1980) The role of wood debris in forests and streams. In: Waring RH (ed) Forests: fresh perspectives from ecosystem analysis. Proc 40th Biology Colloq, 1979, Oregon State University Press, Corvallis

Tubaki K (1958) Studies of the Japanese hyphomycetes V. Leaf and stem group with a discussion of the classification of hyphomycetes and their perfect stages. J Hattori Bot Lab 20:142–244

Tubaki K (1966) An undescribed species of *Hymenoscyphus*, a perfect stage of *Varicosporium*. Trans Br Mycol Soc 49:345–349

Webber JF, Hedger JN (1986) Comparisons of interactions between *Ceratocystis ulmi* and elm bark saprobes in vitro and in vivo. Trans Br Mycol Soc 86:93–101

Webster J (1959) *Nectria lugdunensis* sp. nov., the perfect state of *Heliscus lugdunensis*. Trans Br Mycol Soc 42:322–327

Webster J, Descals E (1981) Morphology, distribution, and ecology of conidial fungi in freshwater habitats. In: Cole GC, Kendrick B (eds), Biology of conidial fungi, vol 1. Academic Press, New York

Wicklow DT (1981) Interference competition and the organization of fungal communities. In: Wicklow DT, Carroll GC (eds) The fungal community, its organization and role in the ecosystem. Marcel Dekker, New York, pp 351–375

Willoughby GL, Archer JF (1973) The fungal spora of a freshwater stream and its colonization on wood. Freshwater Biol 3:219–239

Zare-Maivan H, Shearer CA (1988a) Extracellular enzyme production and cell wall degradation by freshwater lignicolous fungi. Mycologia 80:365–375

Zare-Maivan H, Shearer CA (1988b) Wood decay activity and cellulase production by freshwater lignicolous fungi. Int Biodeterior Bull 24:459–474

Zemek J, Marvanova L, Kuniak L, Kadlecikova B (1985) Hydrolytic enzymes in aquatic hyphomycetes. Folia Microbiol 30:363–372

5 Anamorph-Teleomorph Relationships

John Webster

5.1 Introduction

About 300 species of Ingoldian hyphomycetes are now known (Descals et al. 1993). They grow mostly on leaves and twigs in rapidly flowing freshwater streams, but some are found in lakes, in stagnant water and some in terrestrial habitats but dispersed in water (see Chap. 2). It is probably best to regard some of them as amphibious fungi. Their amphibious nature is illustrated by the fact that although most of them develop conidia freely underwater, some will continue to sporulate in air on twigs or leaves removed from water, and some will also sporulate in culture on agar without free water. Some can survive for months on air-dried leaves removed from streams (Sanders and Webster 1978) (see Chaps. 2, 10). Tree trunks, branches, twigs and leaf petioles exposed to air by receding water levels or stranded on river banks and held in a humid environment frequently bear ascocarps and occasionally basidiocarps. A small proportion of these have been shown to be teleomorphs of Ingoldian fungi. They do not need to be submerged in water in order to discharge their spores, but (with the exception of *Camptobasidium*) do so freely in air. These air-borne spores permit dispersal over large distances and this may explain the paradox of the worldwide distribution of freshwater fungi with passively dispersed conidia. It may also help to explain why such fungi are to be found near the sources of freshwater streams. The opportunities which sexual reproduction provide for gene recombination and selection must also be important for the evolutionary success of fungi with teleomorphs. Over 20 Ingoldian fungi with ascomycetous teleomorphs are now known (Table 5.1) and four with Basidiomycetes (Table 5.2). Some tetraradiate conidia previously described from freshwater foam have been connected with members of the Zygomycete order Entomophthorales and parasitize certain insects with aquatic larval stages.

Table 5.1. Ascomycete teleomorphs of Ingoldian hyphomycetes

Anamorph	Teleomorph	Reference
Branched conidia		
Actinospora megalospora	*Miladina lecithina* (Pezizales)	Descals and Webster (1978)
Anavirga dendromorpha	*Apostemidium torrenticola* (Helotiales)	Hamad and Webster (1987)
Articulospora tetracladia	*Hymenoscyphus tetracladius* (Helotiales)	Abdullah et al. (1981)
Clavariopsis aquatica	*Massarina* sp. (Dothideales)	Webster and Descals (1979)
Geniculospora grandis	*H. africanus* (Helotiales)	Descals et al. (1984)
Heliscus lugdunensis	*Nectria lugdunensis* (Hypocreales)	Webster (1959b)
Tricladium chaetocladium	*Hydrocina chaetocladia* (Helotiales)	Webster et al. (1990)
T. splendens	*Hym. splendens* (Helotiales)	Abdullah et al. (1981)
Varicosporium sp.	*Hym. varicosporoides* (Helotiales)	Tubaki (1966)
Sigmoid conidia		
Anguillospora crassa	*Mollisia uda* (Helotiales)	Webster (1961)
A. furtiva	*Rutstroemia* sp. (Helotiales)	Webster and Descals (1979)
A. fustiformis	*Rutstroemia* sp. (Helotiales)	Webster and Descals (unpublished)
A. longissima	*Massarina* sp. (Dothideales)	Willoughby and Archer (1973)
A. rosea	*Orbilia* sp. (? Helotiales)	Webster and Descals (1979)
Anguillospora sp.	*H. imberbis* (Helotiales)	Webster and Descals (unpublished)
Anguillospora sp.	*Loramyces juncicola* (Helotiales)	Digby and Goos (1987)
Flagellospora curta	*Nectria* sp. (Hypocreales)	Webster (unpublished)
F. penicillioides	*N. penicillioides* (Hypocreales)	Ranzoni (1956)
Filosporella sp.	*Mollisia* sp. (Helotiales)	Webster and Descals (1979)
Conidia of other shape		
Cylindrocarpon ianthothele	*N. mammoidea* (Hypocreales)	Booth (1959)
Dimorphospora foliicola	*H. foliicolus* (Helotiales)	Abdullah et al. (1981)
Tumularia aquatica	*M. aquatica* (Dothideales)	Webster (1965)

5.2 Techniques

Reynolds (1987) has discussed various ways of proving the anamorphic connection: by association, evidence of organic connection between anamorph and teleomorph, or by cultural means, i.e. obtaining an anamorph in a culture derived from the spore of a teleomorph or the converse. Evidence derived merely from association of one morph with another is unacceptable as proof, but constant association may provide a valuable clue

Table 5.2. Basidiomycete teleomorphs of Ingoldian hyphomycetes

Anamorph	Teleomorph	Reference
Taeniospora gracilis	*Leptosporomyces galzinii* (Corticiaceae)	Nawawi et al. (1977)
T. gracilis var. *enecta*	*Fibulomyces* sp. (Corticiaceae)	Marvanová and Stalpers (1987)
T. descalsii	*F. crucelliger* (Corticiaceae)	Marvanová and Stalpers (1987)
Ingoldiella hamata	*Sistotrema hamatum* (Corticiaceae)	Nawawi and Webster (1982)
Crucella subtilis	*Camptobasidium hydrophilum* (Atractiellales)	Marvanová and Suberkropp (1990)

which could prompt more critical testing. Direct connections between anamorph and teleomorph have not been demonstrated for Ingoldian fungi, but synanamorphy, i.e. the existence of two morphologically distinct anamorphs, has occasionally been demonstrated by this means, for example the attachment of a phialidic microconidial synanamorph to the macroconidium of *Casaresia sphagnorum* (Webster and Descals 1975) or *Anavirga dendromorpha* (Hamad and Webster 1987). Since most Ingoldian fungi grow readily in pure culture the demonstration by cultural means is the method of choice. Care should be taken to isolate from a single conidium, ascospore (or ascus) or basidiospore. Evidence based on crude cultures started from ascocarp tissue is unreliable because cases have been reported where apothecial tissue has been pervaded by the mycelium of an unrelated Ingoldian or other fungus.

Webster and Descals (1979) have described techniques for establishing the pleomorphy of Ingoldian fungi by cultural means. The essential steps are the isolation and germination of conidia, ascospores or basidiospores, preferably on dilute media with added antibiotic, transfer of hyphal tips derived from single germinated spores to suitable media, usually Malt extract agar (MEA) or Potato dextrose agar (PDA) and the incubation of such cultures at low temperatures in diffuse light over long periods, often several weeks or months. Conidium formation may be induced by floating culture pieces in static sterile water, or in sterile water aerated by compressed air, or by placing them in a flow cell irrigated by sterile water. Ascocarp formation may be induced by placing culture pieces in sterile water so that the upper surface of the culture is at water level. Prolonged incubation may be necessary before ascocarps appear. If a fungus is homothallic the chances of obtaining teleomorphs in culture are obviously greater than in the case of a heterothallic fungus.

Conidial material suitable for isolation may be from suspensions in foam or from spores developing from submerged leaves, twigs, etc. Ascocarp or basidiocarp material may be collected on woody host material in the field, or may develop on randomly collected twigs incubated in moist chambers, e.g. plastic lunch boxes, in the cool for a few weeks (Willoughby and Archer 1973). A useful technique for ascocarps is to allow ascospores to be projected upwards onto agar. Since discharge of basidiospores occurs over a shorter range than ascospores it is preferable to arrange pieces of basidiocarp attached to Petri-dish lids so that projected spores can fall onto agar lying beneath.

5.3 Nomenclature

The existence of pleomorphic states, and sometimes of pleoanamorphic states in one fungus raises problems of nomenclature which have been addressed by Hennebert (1987). Whilst many mycologists would prefer the ideal of "one species one name" strongly advocated by Mason (1937), in practice a kind of dual nomenclature has been sanctioned with separate names for different morphs. I prefer to use a cross-reference name method in which, for example, one may refer to the *Anguillospora crassa* state of *Mollisia uda*, or the converse, both names having separate epithets and validity. Since the Ingoldian hyphomycetes are best known by the names of their macroconidial anamorphs I shall generally use these names.

5.4 Synanamorphs

Table 5.1 lists only the macroconidial anamorphs of Ingoldian fungi, but, as indicated, in several cases a synanamorph is also known. There are essentially two types of synanamorph: microconidia which are capable of germination and probably have an asexual role, and spermatia which do not germinate and which possibly have a sexual function. Microconidia capable of germination are known for *Heliscus lugdunensis* and *Cylindrocarpon ianthothele*, both of which are conidial states of *Nectria* spp. In both cases the microconidia are phialoconidia as indeed are the corresponding macroconidia. As pointed out by Samuels and Seifert (1987), "phialidic conidiogenesis is the norm in the Hypocreales". Spermatia, which are generally colourless, minute, globose, ovoid or bacilliform also show phialidic conidiogenesis, often developing from phialides with a long cylindrical collarette. The conidiophores may be mononematous or occasionally develop inside pycnidia (spermogonia). Pycnidial spermatia have been described for *Clavariopsis aquatica*, *Anguillospora longissima* and *Tumularia aquatica* and it may be significant that all three have teleomorphs belonging to the Loculoascomycete genus *Massarina* (Webster and Descals 1979). Mononematous conidiophores with phialidic spermatia are known for several fungi, e.g. *A. longissima*, *Filosporella* sp. (Webster and Descals 1979), *Tricladium splendens* (Abdullah et al. 1981), and *T. chaetocladium* (Webster et al. 1990), see Fig. 5.1C. The proof that these minute phialoconidia are spermatia, i.e. that they

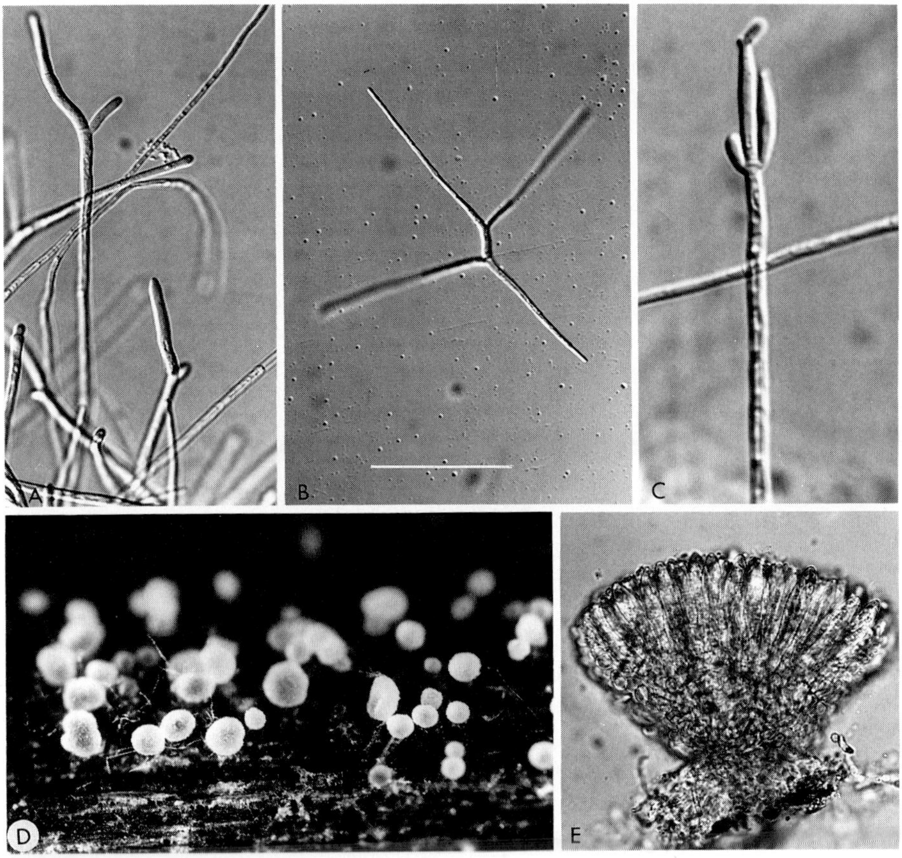

Fig. 5.1A–E. *Tricladium chaetocladium*. **A** Macroconidiophores and developing macroconidia; **B** detached macroconidium; **C** microconidial phialidic synanamorph; **D** apothecial teleomorph (*Hydrocina chaetocladia*) growing on a twig; **E** lateral view of an apothecium. Scale *bar* **A** = 25 µm; **B,E** = 100 µm, **D** = 700 µm

have a sexual function, has not yet been made. *Tricladium splendens* is known to be heterothallic and only forms its *Hymenoscyphus* state in cultures derived from more than one ascospore. It would be interesting to transfer "spermatia" from one mating type to another to see if apothecia would develop.

5.5 Entomophthorales with Branched Conidia

Along with branched and sigmoid conidia of well-known Ingoldian hyphomycetes, large tetraradiate conidia of varied shape with refringent granular contents are often found in freshwater foam. Ingold (1975b) sug-

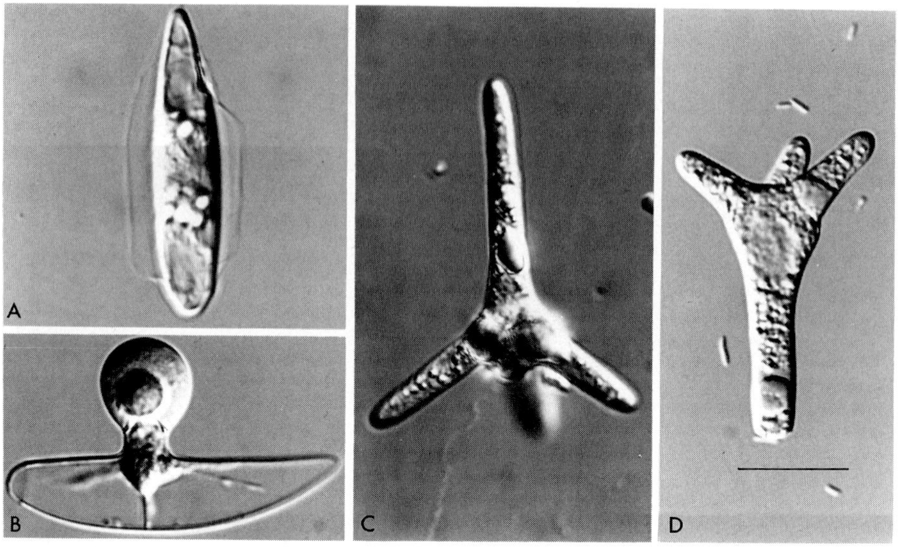

Fig. 5.2A–D. *Erynia conica* and its four synanamorphs. **A** Primary cornute aerial conidium which has developed from a conidiophore projecting into the air from a dead insect showing its bitunicate wall; **B** secondary globose conidium which has developed from a primary conidium germinating in air; **C** secondary stellate conidium which has developed from a primary conidium germinating under water; **D** primary coronate conidium which has developed directly from a conidiophore projecting into water from a submerged infected insect. Scale *bar* = 20 μm

gested that they might be rhizopods. Chance discovery of organic connection between such branched conidia and conidia of a different, cornute shape led to the suggestion that the tetraradiate conidia belonged to a member of the Entomophthorales (Webster et al. 1978). This was confirmed when infected insects were found along stream margins and by weirs and waterfalls (Descals et al. 1981; Descals and Webster 1984; Webster 1987). Several species are known to form tetraradiate conidia, e.g. *Erynia conica*, *E. rhizospora*, *E. plecopteri*, *Entomophthora* cf *thaxteri* and *Ent. gigantea*, but further connections probably also exist with other species. The most versatile is *Er. conica* which is especially common on female *Simulium* spp. This fungus has four distinctive conidial types (Fig. 5.2), two formed aerially (a primary cornute and a secondary globose conidium) and two underwater. The two kinds of conidia formed underwater are tetraradiate but differ from each other in their ontogeny. The primary *coronate* conidia develop at the tips of conidiophores projecting from the dead insect host into water, whilst the secondary *stellate* conidia develop underwater, usually from one of the aerial conidia. Germination often occurs not by germ tube but by the development of secondary conidia, and of the 16 (4 × 4) possible inter-conversions between the four conidial types, 12 have been observed (Webster 1987). Curiously, although it is not possible to prove a negative, there is no evidence that the primary cornute, primary coronate or secondary stellate conidia ever cause infection, a function which appears to be confined to the secondary globose conidium (Hywel-Jones and Webster 1986). The teleomorph of *Er. conica* is in the form of globose zygospores (or azygospores) embedded in a network of brown hyphae in the cadaver of an infected host.

5.6 Ascomycete Teleomorphs

The first connection between an Ascomycete and an Ingoldian fungus was discovered by Ranzoni (1956) when perithecia of a *Nectria*, *N. penicillioides*, appeared in cultures derived from phialoconidia of *Flagellospora penicillioides*. Single ascospore cultures produced the *Flagellospora* again. Perithecia developed only when certain single ascospore isolates were paired, suggesting that the fungus is heterothallic. Since then, as Table 5.1 shows, further connections have been made, either as a result of teleomorphs arising in cultures derived from conidia, or the discovery of conidia in cultures started from ascospores. An apparently undescribed species of *Flagellospora*, provisionally named *F. curta*, has recently been obtained in cultures started from *Nectria* ascospores (Webster, unpubl.). *Nectria* teleomorphs are also known for *Heliscus lugdunensis* and *Cylindrocarpon ianthothele* (Fig. 5.3). Twigs such as those of oak (*Quercus*), alder (*Alnus*), and other deciduous trees when submerged in streams quickly become colonized by *Heliscus*, *Cylindrocarpon* and *Fusarium* forming conspicuous sporodochium-like pustules. When incubated under moist cool conditions the perithecia of their related *Nectria* often develop around the conidial pustule (Shearer and Webster 1991).

The list of Ascomycete genera in Table 5.1 shows that certain genera such as *Nectria*, *Massarina* and *Hymenoscyphus* include teleomorphs of several Ingoldian anamorphs. In addition to the *Cylindrocarpon*, *Heliscus*, *Flagellospora* and *Fusarium* anamorphs of *Nectria* associated with aquatic environments, there is a very wide range of terrestrial anamorphs including *Gliocladium*, *Penicillifer*, *Sesquicillium*, *Tubercularia* and *Myrothecium* (Samuels and Seifert 1987; Samuels 1989; Rossman 1983). In terms of habitat and conidial morphology *Nectria* and its allies thus show remarkable evolutionary divergence.

The Loculoascomycete genus *Massarina* has forms with macroconidia ascribed to three anamorph genera viz *Anguillospora longissima* (sigmoid), *Clavariopsis aquatica* (tetraradiate) and *Tumularia aquatica* (pyriform). Although the morphology of the macroconidia differs markedly (Fig. 5.4) there are similarities between these three species all of which have dark-coloured mycelia, pycnidial "spermatia" and percurrent proliferation of the conidiophores. Pseudothecia of *Massarina* spp. occur frequently on the wood of branches of oak, alder and other trees which have been in a

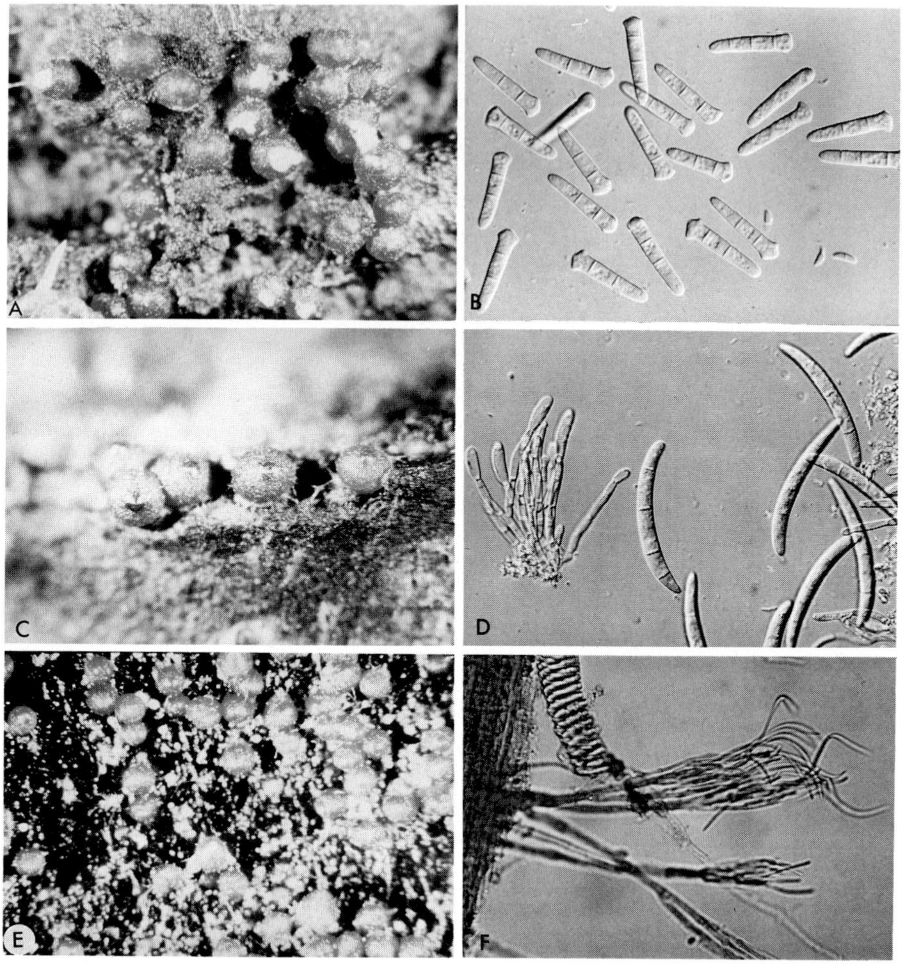

Fig. 5.3A–F. *Nectria* species with aquatic conidia. **A** *Nectria lugdunensis* perithecia on bark from a previously submerged twig; **B** *Heliscus lugdunensis* macro- and micro-conidia; **C** *Nectria mammoidea* perithecia on bark from a previously submerged twig; **D** *Cylindrocarpon ianthothele* macroconidial state of *N. mammoidea*; **E** *Nectria penicillioides* perithecia in culture derived from *Flagellospora* phialoconidia; **F** *Flagellospora penicillioides* phialoconidiophores and phialoconidia-stained preparation from a submerged tropical leaf

river for many months. It is likely that ascospores are dispersed into water, although they can also be discharged into air. In all species there is a conspicuous mucilaginous epispore around the ascospore which may play a role in attachment (Fig. 5.5C). Other species of *Massarina* which apparently do not form conidia in culture also form pseudothecia on submerged wood. Scheuer (1990) has shown that *Tetraploa aristata* a dematiaceous

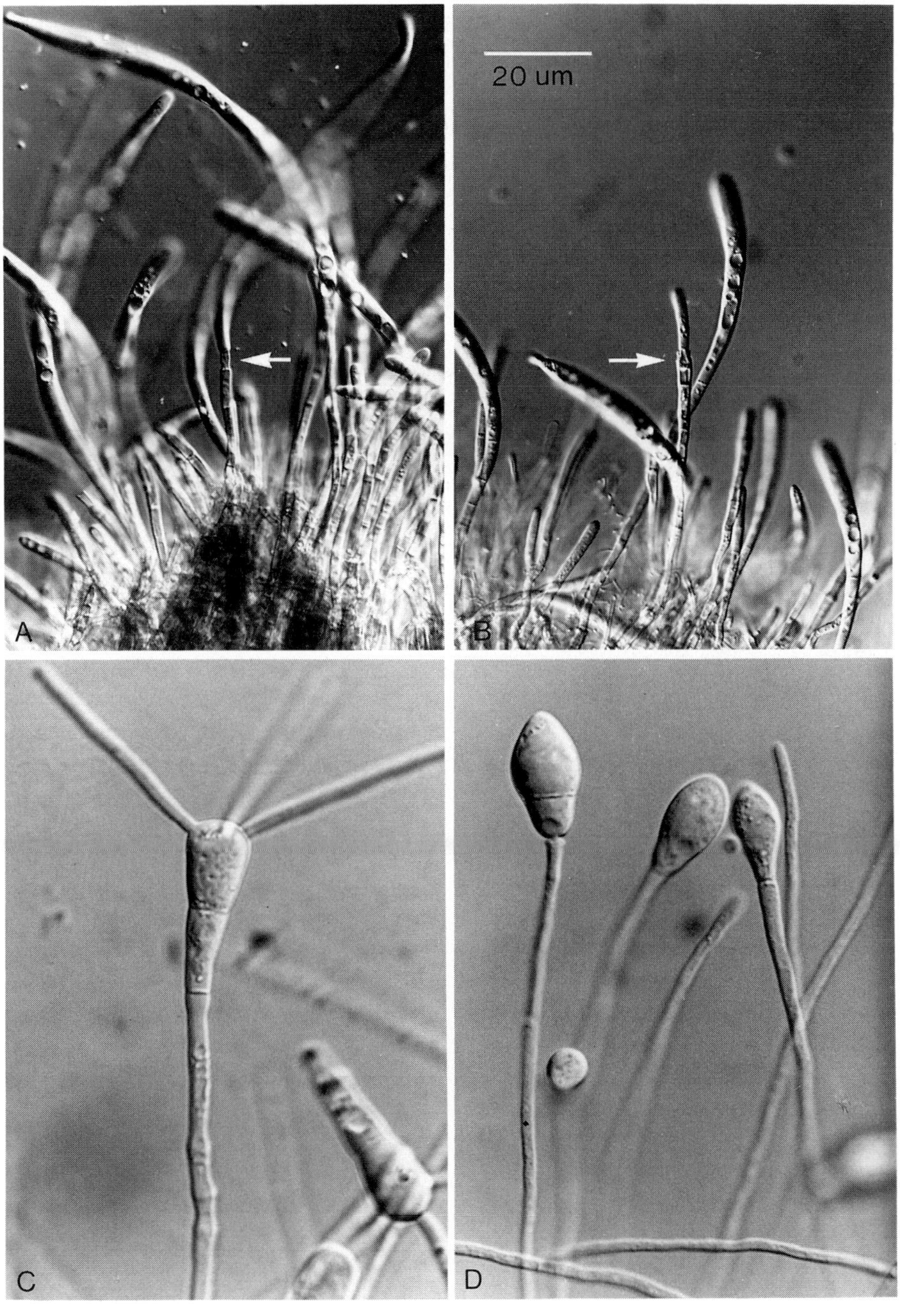

Fig. 5.4A–D. Conidial states of *Massarina* spp. **A** *Anguillospora longissima* macro-conidiophores and macroconidia. The separating cell at the base of a macro-conidium is *arrowed*; **B** *A. longissima* macronidiophores. Percurrent proliferation through a ruptured separating cell is *arrowed*; **C** *Clavariopsis aquatica* macro-conidiophore and conidium, **D** *Tumularia aquatica* macro-conidiophores and macroconidia. All to same scale

Fig. 5.5A–C. *Massarina aquatica*. **A** V.S. pseudothecium; **B** Squash preparation showing asci and pseudoparaphyses; **C** ascospore showing mucilaginous sheath. Scale *bar* **A** = 1 mm, **B** = 100 μm, **C** = 50 μm

hyphomycete with large dark tetraradiate conidia which is common on moist decaying monocotyledonous herbs such as grasses and sedges has a *Massarina* teleomorph, *M. tetraploa*. It is of interest to speculate on the selection pressures which have conserved a rather uniform pseudothecial and ascospore morphology and have led to the evolution of such different kinds of conidia, presumably with different strategies for attachment (Webster 1959a; Webster and Davey 1984; Read 1990). Assuming that these aquatic species of *Massarina* are congeneric with those growing on terrestrial substrata the problem becomes even more intriguing because conidia have been reported from terrestrial species ascribed to pycnidial anamorph genera such as *Coniothyrium*, *Ceratophoma* and *Microsphaeropsis* (Sivanesan 1984).

The discomycete genus *Hymenoscyphus* has Ingoldian anamorphs in *Articulospora*, *Geniculospora*, *Tricladium*, *Varicosporium*, *Anguillospora* and *Dimorphospora*. Apothecia develop in terrestrial situations on fallen leaves, cupules, herbaceous stems, etc. About 100 species are known, but relatively few have been grown in culture. *Hymenoscyphus paradoxus* has the aeroaquatic anamorph *Helicodendron paradoxum* (Fisher and Webster 1983). *Hymenoscyphus caudatus* has an *Idriella* anamorph with narrowly fusiform conidia (Kimbrough and Atkinson 1972).

The Discomycete genus *Mollisia* includes about 100 species. Culture studies have shown that two species have anamorphs of the *Anguillospora* and *Filosporella* type. It is likely that further connections will be made with Mollisioid fungi. Aeroaquatic anamorphs are also known: *Helicodendron giganteum* (Fisher and Webster 1983) and *Spirosphaera floriformis* (Abdullah 1980).

The anamorph genus *Anguillospora* is clearly heterogeneous. The type species is *A. longissima*, with a *Massarina* (Loculoascomycete) teleomorph. The conidia are holoblastic. Conidial secession is rhexolytic: a specialized separating cell at the base of the conidiophore loses its cytoplasmic contents and collapses. The cylindrical wall of the separating cell breaks circumscissilely near the middle, leaving a tubular collar at the base of the conidium and at the apex of the conidiophore (see Fig. 5.4B). Percurrent growth of the conidiophore through the collar is followed by the delimitation of a further holoblastic conidium at a higher level. Interpreted in this narrow sense it is clear that many other taxa classified in *Anguillospora* do not develop in the same way. *Anguillospora furtiva*, which has conidia which closely resemble those of *A. longissima* in dimensions and septation, shows schizolytic conidial separation. Its teleomorph, *Rutstroemía* sensu Korf, sp. belongs to an unrelated group of Ascomycetes, the Helotiales. Other *Anguillospora* teleomorphs belong to *Hymenoscyphus, Mollisia* and to *Loramyces*. It is of interest that the marine Ascomycete *Lindra obtusa* has *Anguillospora marina* as its anamorph (Nakagiri and Tubaki 1987).

The form genus *Tricladium* is likewise heterogeneous. The type species *T. splendens* has a *Hymenoscyphus* teleomorph whilst *T. chaetocladium* has apothecia belonging to the monotypic genus *Hydrocina*. As Ingold (1975a) has pointed out, *Tricladium* is by far the largest genus of aquatic hyphomycetes and is in need of critical revision. Further cultural studies should help in such revision.

5.7 Basidiomycete Teleomorphs

Although the majority of Ingoldian fungi are believed to be related to Ascomycetes, a minority are Basidiomycetes. Evidence for this is in the presence of clamp connections on mycelial or conidial septa or on conidial germ tubes indicating dikaryotic propagules, or the possession of dolipore septa, a characteristic feature of certain groups of Basidiomycetes (Nawawi 1985; Marvanová and Bandoni 1987). In a few cases the connection between anamorphs and teleomorphs has been established in culture (see Table 5.2).

Marvanová (1977) gave the epithet *Taeniospora gracilis* to a fungus with tetraradiate conidia having a curved main axis from which two laterals develop on the convex side. Between the two laterals is a clamp connection (Fig. 5.6B). Cultures derived from such conidia formed athelioid basidiocarps with four spored basidia (Nawawi et al. 1977). At that time the teleomorph was identified as *Leptosporomyces galzinii*, a fungus whose basidiocarps had previously been collected on wood or bark of conifers, frondose trees and on soil (Jülich 1972). Later more critical studies by Marvanová and Stalpers (1987) established that two distinct intraspecific taxa in *Taeniospora gracilis* could be distinguished. The fungus with athelioid

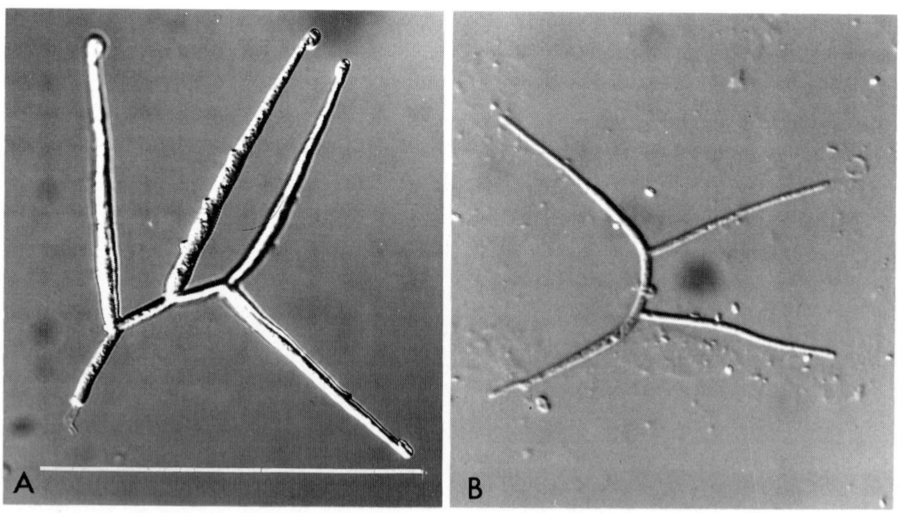

Fig. 5.6. **A** *Ingoldiella hamata*: dikayotic conidium showing clamp connections at the septa. **B** *Taeniospora gracilis*: dikaryotic conidium showing clamp connections at the septum between the lateral arms. Scale *bar* **A** = 100 μm, **B** = 70 μm

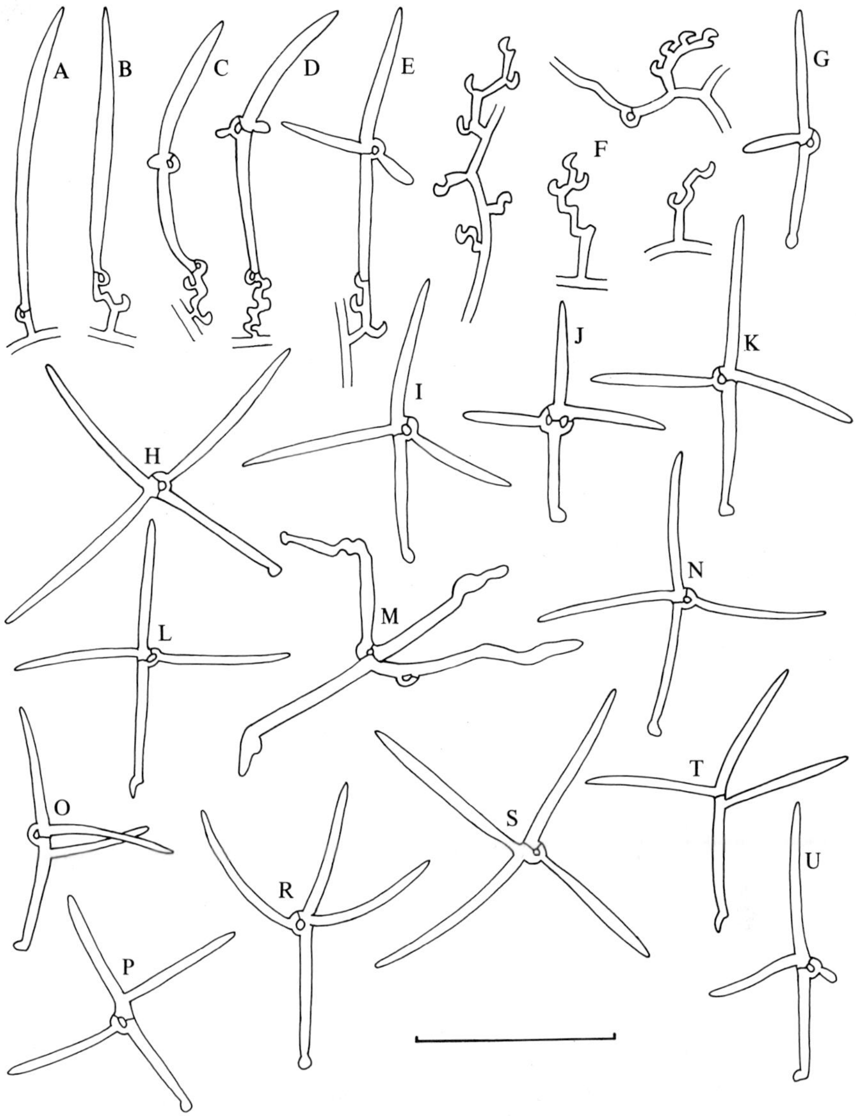

Fig. 5.7A–U. *Taeniospora descalsii*. **A–E** Developing conidia; **F** spent conidiophores; **G–U** Free conidia; **M** germinating conidium; **G,J,O** aberrant conidia; **T** clampless conidium from foam. Scale *bar* = 20 µm (Marvanová and Stalpers 1987)

basidiocarps studied by Nawawi et al. was described as *T. gracilis* var. *enecta* and, because doubt had been attached to the name of its teleomorph, this was ascribed to *Fibulomyces* sp. Another variety of *Taeniospora*, *T. gracilis* var *gracilis*, was distinguished for which no teleomorph has yet been observed. A second species, *T. descalsii*, was described (Fig. 5.7) and this developed a teleomorph in culture, *Fibulomyces crucelliger* (Fig. 5.8).

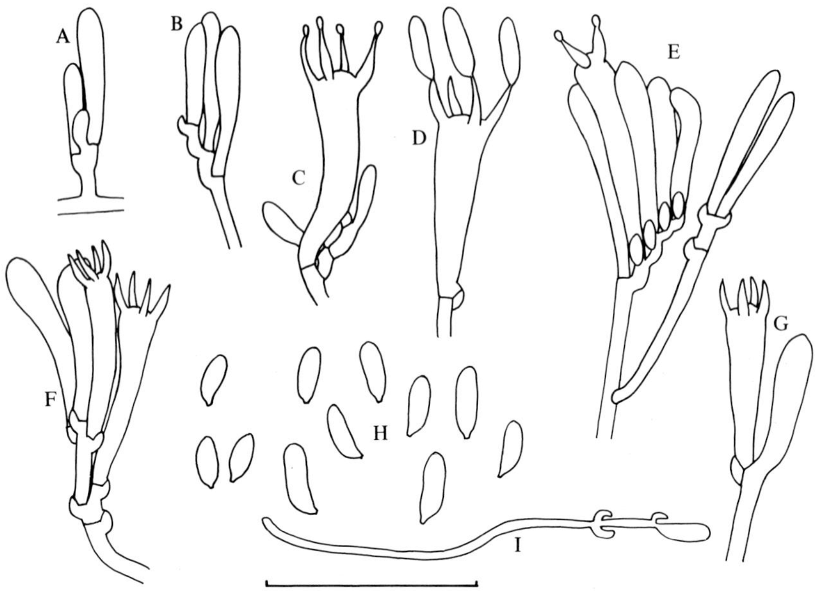

Fig. 5.8A–I. *Fibulomyces crucelliger*, the teleomorph of *Taeniospora descalsii*. **A–G** Basidia in various stages of development; **H** basidiospores; **I** germinating basidiospore with incomplete clamps. Scale *bar* = 20 μm (Marvanová and Stalpers 1987)

Ingoldiella hamata, a fungus known mostly from warmer regions, forms very large dikaryotic multiseptate conidia with a main axis and two to four laterals. The tips of the main axis and of branches are hooked, and there is a clamp connection at each septum (see Fig. 5.6A). Nawawi and Webster (1982) showed that its teleomorph was a *Sistotrema* with basidia bearing up to six basidiospores. The basidiospores germinate to form monokaryotic mycelia which develop monokaryotic propagules similar in form to the dikaryotic propagules, but lacking clamps. *Sistotrema* and *Fibulomyces* both belong to the Corticiaceae and have holobasidia.

Marvanová and Suberkropp (1990) have established the connection between the heterobasidiomycete *Camptobasidium hydrophilum* and an anamorph with tetraradiate conidia which are clamped at the base of the conidiogenous cell (Fig. 5.9). The basidia are curved and transversely septate (Fig. 5.10). Each cell of the four celled basidium may bear several basidiospores which remain clustered together. The mycelial septa lack dolipores. The fungus fails to grow on sterile leaf material, but when grown with other aquatic hyphomycetes, the *Camptobasidium* hyphae coiled around the hyphae of its partners, and it was suggested that *Camptobasidium* may be an unspecialized biotrophic mycoparasite. In this respect it is similar to another conidial aquatic Basidiomycete *Naiadella fluitans* (Marvanová and Bandoni 1987) which forms haustorial branches similar to those of

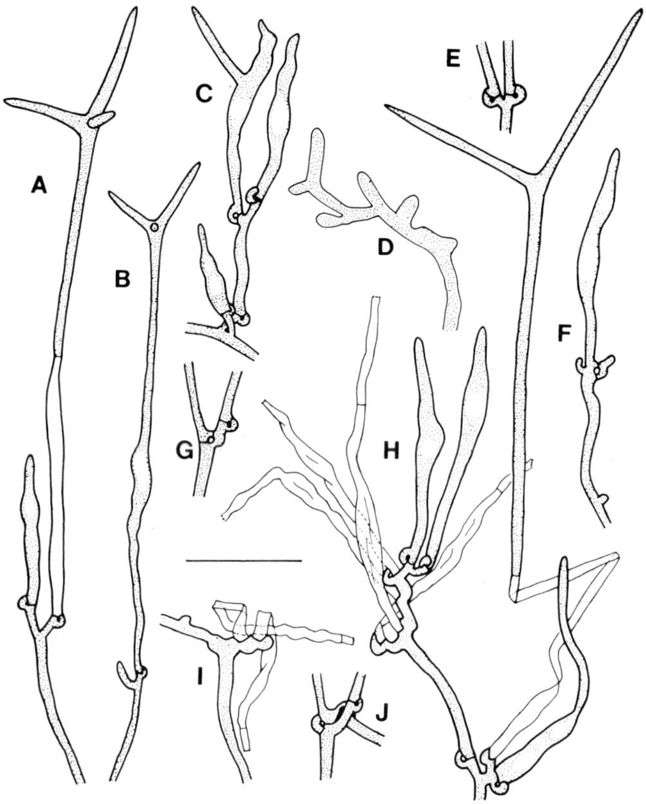

Fig. 5.9A–J. *Crucella subtilis* state of *Camptobasidium hydrophilum*. **A,B** Conidia on conidiogenous cells; **C,F** developing conidiogenous cells; **D** typical branching of the tips of hyphae; **H** conidiophore with conidiogenous structures at various stages of development; **I** spent conidiogenous cells; **E,G,J** various kinds of mycelial branch origin. Scale *bar* = 20 μm (Marvanová and Suberkropp 1990)

mycoparasitic Tremellales. *Naiadella fluitans* develops dikaryotic propagules from a clamped mycelium. So far no teleomorph has been seen, and it would be of interest to cultivate it along with a suitable host fungus to see if the basidiomata would develop to permit the putative classification in the Tremellales to be confirmed. This idea should also be extended to some of the other known dikaryotic Basidiomycete conidial forms which have so far not yielded evidence of teleomorphs.

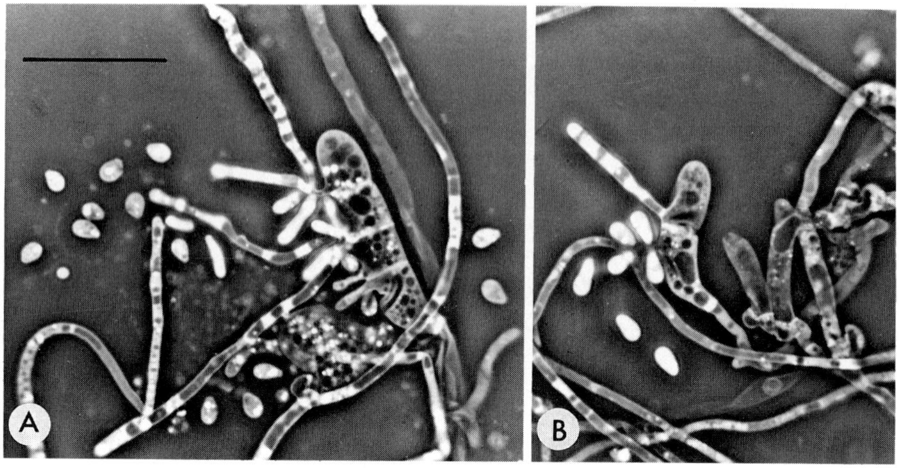

Fig. 5.10A,B. *Camptobasidium hydrophilum.* **A,B** Basidia with basidiospores from submerged culture. Scale *bar* = 20 µm (Marvanová and Suberkropp 1990)

5.8 Discussion

It is not surprising, given the spectacular and unusual shapes of conidia of Ingoldian fungi, that attention should have been focussed on these anamorphic states in relation to development, classification, ecology, trapping and other features of their biology to the neglect of the holomorph, i.e. the whole fungus. Knowledge of their teleomorphs obviously provides information about their relationships with other groups of fungi, some of which have terrestrial representatives. It also adds emphasis to the view that the two most common variants of conidial shape, the tetraradiate (or branched) and the sigmoid, represent examples of convergent evolution in aquatic habitats, possibly as an aid to underwater trapping or as an adaption to dispersal in water films (Ingold 1975b; Webster 1987). In most Ingoldian fungi from freshwater diversity in propagule morphology is centred on the anamorph with comparatively little variation in ascospore or basidiospore morphology. In contrast, in many marine fungi there is great diversity in the ornamentation of ascospore appendages (Kohlmeyer and Kohlmeyer 1979; Kohlmeyer 1986). Relatively few marine Ascomycetes appear to have branched or sigmoid conidia (Shearer 1986). However, marine Basidiomycetes are known with tetraradiate basidiospores e.g. *Nia*, *Digitatispora*, or sigmoid basidiospores formed within pycnidial basidiomata as in *Mycaureola.* These striking differences between the morphology of ascospores and basidiospores of freshwater fungi on the one hand and marine fungi on the other suggest that in freshwater fungi the large, branched or sigmoid conidia are adapted to dispersal under water or in water films, whilst their smaller conventionally shaped ascospores and basidiospores are mostly disseminated in air. In most marine fungi the ascospores and basidiospores are actually released underwater, and are apparently adapted to being trapped on underwater substrata.

It is clear that several of the anamorph genera of Ingoldian fungi are artificial, heterogeneous assemblages. As Samuels and Seifert (1987) have pointed out, "unfortunately names of anamorph fungi do not imply more than morphological similarity".

The earlier studies on Ingoldian fungi were based mostly on the examination of allochthonous tree leaves as substrata. Whilst conidia are formed in abundance on submerged leaf material, the available evidence is that with a few exceptions (*Hymenoscyphus tetracladius* and *H. africanus*) teleomorphs seem not to occur on leaves, but grow on wood but prolonged incubation of leaf material has rarely been attempted. Possibly this is because woody substrata provide larger amounts of nutrient to support the development of ascomata, or that, being more enduring and less exposed to consumption by aquatic invertebrates, they provide a sufficiently long-lasting substratum to permit ascocarp development. Where teleomorphs have developed in cultures derived from conidia in foam, the normal substratum for the teleomorph is not known, and further search on exposed woody material or on the emergent portions of instream or riparian vegetation might be rewarding (see also Chap. 4).

Acknowledgements. It is a pleasure to acknowledge help and valuable discussion with my former colleagues and research students Dr. E. Descals, Prof. A. Nawawi, Dr. S.K. Abdullah, Dr. P.J. Sanders, Dr. P.J. Fisher, Dr. L. Marvanová, Dr. Ch. Scheuer, Prof. C. Shearer and Mrs Om Kalthoum Khattab. I am particularly grateful to Dr. Marvanová for permission to use drawings and photographs.

References

Abdullah SK (1980) British aero-aquatic fungi. PhD Thesis, Exeter University

Abdullah SK, Descals E, Webster J (1981) Teleomorphs of three aquatic Hyphomycetes. Trans Br Mycol Soc 77:475–483

Booth C (1959) Studies of pyrenomycetes IV. *Nectria* (part 1). Commonw Mycol Inst Mycol Pap 73:1–115

Descals E, Webster J (1978) *Miladina lecithina* (Pezizales), the ascigerous state of *Actinospora megalospora*. Trans Br Mycol Soc 70:466–472

Descals E, Webster J (1984) Branched aquatic conidia in *Erynia* and *Entomophthora sensu lato*. Trans Br Mycol Soc 83:669–682

Descals E, Webster J, Ladle M Bass JAB (1981) Variations in asexual reproduction in species of *Entomophthora* in aquatic insects. Trans Br Mycol Soc 77:85–102

Descals E, Fisher PJ, Webster J (1984) The *Hymenoscyphus* teleomorph of *Geniculospora grandis*. Trans Br Mycol Soc 83:541–546

Descals E, Marvanová L, Webster J (1993) A monograph of Ingoldian hyphomycetes (in preparation)

Digby S, Goos RD (1987) Morphology, development and taxonomy of *Loramyces*. Mycologia 79:821–831

Fisher PJ, Webster J (1983) The teleomorphs of *Helicodendron giganteum* and *H. paradoxum*. Trans Br Mycol Soc 81:656–659
Hamad SR, Webster J (1987) *Anavirga dendromorpha*, anamorph of *Apostemidium torrenticola*. Sydowia 40:60–64
Hennebert GL (1987) Pleoanamorphy and its nomenclatural problem. In: Sugiyama J (ed) Pleomorphic fungi: the diversity and its taxonomic implications. Kodansha, Tokyo, pp 29–56
Hywel-Jones NL, Webster J (1986) Mode of infection of *Simulium* by *Erynia conica*. Trans Br Mycol Soc 87:381–387
Ingold CT (1975a) An illustrated guide to aquatic and water borne Hyphomycetes (Fungi Imperfecti) with notes on their biology. Freshwater Biol Assoc Sci Publ No 30. Ambleside, Cumbria, England
Ingold CT (1975b) Convergent evolution in aquatic fungi: the tetraradiate spore. Biol J Linn Soc 7:1–25
Jülich W (1972) Monographie der Athelieae (Corticiaceae, Basidiomycetes). Willdenowia Beih 7:283
Kimbrough JW, Atkinson M (1972) Cultural features and imperfect stage of *Hymenoscyphus caudatus*. Am J Bot 59:165–171
Kohlmeyer J (1986) Taxonomic studies of marine Ascomycotina. In: Moss ST (ed) The biology of marine fungi. Cambridge University Press, pp 199–210
Kohlmeyer J, Kohlmeyer E (1979) Marine mycology: the higher fungi. Academic Press, New York
Marvanová L (1977) *Taeniospora gracilis* gen. et sp. nov. Trans Br Mycol Soc 69:146–148
Marvanová L, Bandoni RJ (1987) *Naiadella fluitans* gen. et sp. nov.: a conidial Basidiomycete. Mycologia 79:578–586
Marvanová L, Stalpers JA (1987) The genus *Taeniospora* and its teleomorphs. Trans Br Mycol Soc 89:489–498
Marvanová L, Suberkropp K (1990) *Camptobasidium hydrophilum* and its anamorph *Crucella subtilis*: a new heterobasidiomycete from streams. Mycologia 82:208–217
Mason EW (1937) Annotated account of fungi received at the Imperial Mycological Institute. List II. Mycol Pap 4:69–99
Nakagiri A, Tubaki K (1987) Pleomorphy in marine fungi: teleomorph-anamorph connections in the Halosphaeriaceae. In: Sugiyama J (ed) Pleomorphic fungi: the diversity and its taxonomic implications, Kodansha, Tokyo, pp 79–101
Nawawi A (1985) Basidiomycetes with branched water borne conidia. Bot J Linn Soc 91:51–60
Nawawi A, Webster J (1982) *Sistotrema hamatum* sp. nov., the teleomorph of *Ingoldiella hamata*. Trans Br Mycol Soc 78:287–291
Nawawi A, Descals E, Webster J (1977) *Leptosporomyces galzinii*, the basidial state of a clamped branched conidium from fresh water. Trans Br Mycol Soc 68:31–36
Ranzoni FV (1956) The perfect stage of *Flagellospora penicillioides*. Am J Bot 43:13–17
Read SJ (1990) Spore attachment in fungi with special reference to freshwater hyphomycetes. PhD Thesis, Portsmouth Polytechnic
Reynolds DR (1987) Proving the anamorphic connection. In: Sugiyama J (ed) Pleomorphic fungi: the diversity and its taxonomic implications. Kodansha, Tokyo, pp 157–180
Rossman AI (1983) The phragmosporous species of *Nectria* and related genera. Commonw Agric Bur Mycol Pap 150:1–164
Samuels GJ (1989) *Nectria* and *Sesquicillium*. Mem NY Bot Gdn 49:266–285
Samuels GJ, Seifert KA (1987) Taxonomic implications of variation among Hypocrealean anamorphs. In: Sugiyama J (ed) Pleomorphic fungi: the diversity and its taxonomic implications. Kodansha, Tokyo, pp 29–56
Sanders PJ, Webster J (1978) Survival of aquatic hyphomycetes in terrestrial situations. Trans Br Mycol Soc 71:231–237

Scheuer Ch (1990) *Massarina tetraploa* sp. nov., the teleomorph of *Tetraploa aristata*. Mycol Res 95:126–128

Shearer CA (1986) The significance of teleomorph/anamorph connections in the classification of marine Ascomycotina. In: Moss ST (ed) The biology of marine fungi. Cambridge University Press, Cambridge, pp 253–262

Shearer CA, Webster J (1991) Aquatic hyphomycete communities in the River Teign IV. Twig colonization. Mycol Res 95:413–420

Sivanesan A (1984) The Bitunicate Ascomycetes and their Anamorphs. Cramer, Vaduz

Tubaki K (1966) An undescribed species of *Hymenoscyphus*, a perfect stage of *Varicosporium*. Trans Br Mycol Soc 49:345–349

Webster J (1959a) Experiments with spores of aquatic hyphomycetes I. Sedimentation and impaction. Ann Bot (Lond) 23:595–611

Webster J (1959b) *Nectria lugdunensis* sp. nov., the perfect stage of *Heliscus lugdunensis*. Trans Br Mycol Soc 42:322–327

Webster J (1961) The *Mollisia* perfect state of *Anguillospora crassa* Ingold. Trans Br Mycol Soc 44:559–564

Webster J (1965) The perfect state of *Pyricularia aquatica*. Trans Br Mycol Soc 48:449–452

Webster J (1987) Convergent evolution and the functional significance of spore shape in aquatic and semi-aquatic fungi. In: Rayner ADM, Brasier CM, Moore D (eds) Evolutionary biology of the fungi. Cambridge University Press, Cambridge

Webster J, Davey RA (1984) Sigmoid conidial shape in aquatic fungi. Trans Br Mycol Soc 83:43–52

Webster J, Descals E (1975) The phialidic state of *Casaresia sphagnorum*. Trans Br Mycol Soc 64:529–531

Webster J, Descals E (1979) The teleomorphs of water-borne Hyphomycetes from fresh water. In: Kendrick WB (ed) The whole fungus, 2. National Museum of Natural Sciences, Ottawa, Canada, pp 419–451

Webster J, Sanders PF, Descals E (1978) Tetraradiate propagules in two species of *Entomophthora*. Trans Br Mycol Soc 70:472–479

Webster J, Scheuer Ch, Om Khaltoum Khattab S (1990) *Hydrocina chaetocladia* gen. et spec. nov., the teleomorph of *Tricladium chaetocladium*. Nova Hedwigia 52:65–72

Willoughby LG, Archer JF (1973) The fungal spora of a stream and its colonization pattern on wood. Freshwater Biol 3:219–239

6 Interactions with Invertebrates

Keller Suberkropp

6.1 Introduction

Woodland streams receive significant quantities of leaf litter and woody debris from the riparian vegetation. This allochthonous organic matter provides much of the carbon and energy for the detritus-based food chains that are common in streams. Initial studies of the interactions between fungi and invertebrates in streams focused on their role in the energy flow and trophic dynamics of such detritus-based food chains (Triska 1970; Kaushik and Hynes 1971). More recently, studies have also emphasized the role that these interactions may play in regulating both fungal (Bärlocher 1982a) and invertebrate (Rossi 1985; Arsuffi and Suberkropp 1989) communities. Most attention has focused on shredders, the functional feeding group of invertebrates with morpho-behavioral adaptations to consume leaf detritus (Cummins 1973), and the effects of leaf-associated microbiota, particularly fungi, on shredder feeding. A variety of shredder taxa have been examined, including both crustaceans and insects (Table 6.1). With respect to fungi, aquatic hyphomycete species have received the most study, but other fungi (e.g. terrestrial hyphomycetes, Oomycetes and Zygomycetes) isolated from leaves in streams have also been examined. Although interactions may occur between fungi and other groups of invertebrates in streams (Anderson and Sedell 1979; Anderson and Cargill 1987), these interactions have not been extensively studied and will not be discussed here.

When leaves enter a stream, they are colonized by stream microorganisms and undergo a series of transformations. In most streams, aquatic hyphomycetes become the dominant microbial group during the early stages of leaf decomposition. During this period, nitrogen and phosphorus contents increase and concentrations of structural polysaccharides decrease (Suberkropp et al. 1976; Meyer 1980; Rosset et al. 1982; Chauvet 1987). Microbial biomass and activity increase to maxima and decline (Suberkropp and Klug 1976; Rosset et al. 1982; Lawson et al. 1984; Findlay and Arsuffi 1989; Suberkropp 1991). Leaves also become softened by the activities of the microbial community (Suberkropp and Klug 1981). Coinciding with these changes, the leaf typically improves as a food source for shredders

Table 6.1. Representative shredder taxa which have been examined in studies of fungal-shredder interactions

Crustacea

Amphipoda
Gammarus fossarum, *G. minus*, *G. pseudolimnaeus*, *G. pulex*, *G. tigrinus*
Isopoda
Asellus aquaticus, *Lirceus* sp., *Proasellus coxalis*

Insecta

Diptera
Tipula abdominalis, *T. caloptera*
Plecoptera
Peltoperla sp., *Pteronarcella badia*, *Pteronarcys proteus*
Trichoptera
Clistoronia magnifica, *Hesperophylax magnus*, *Phylloicus* sp., *Psychoglypha* sp., *Pycnopyche luculenta*, *P. guttifer*

(e.g. Lawson et al. 1984). "Conditioning" is a term that has been used to describe the changing quality of leaf detritus as a food for invertebrates during microbial colonization and transformation (Boling et al. 1975; Bärlocher 1985). As leaf detritus is conditioned, its food quality increases to a maximum, i.e. the detritus becomes "fully conditioned". During the period of time when leaves are becoming fully conditioned, fungi account for the majority of the microbial biomass associated with leaf detritus (Findlay and Arsuffi 1989). Subsequent changes in leaf detritus, e.g. declining microbial biomass and activity, lead to decreases in food quality or a "post-conditioned" phase. The dynamics of conditioning curves are related to chemical and physical characteristics of leaf species and environmental parameters (e.g. temperature) which affect decomposition rates.

A number of factors have been shown to affect the interactions between aquatic hyphomycetes and shredders. Both the fungal and the shredder taxa involved are diverse and the outcome may vary depending on which species are involved. Fungal-shredder interactions are dynamic and change through time as leaves undergo degradation and conditioning. Since shredder feeding usually follows initial fungal growth on leaf litter, properties of the leaf species (i.e. structural attributes, chemical composition, etc.) can affect the nature of the interactions. In addition, characteristics of the habitat (e.g. water chemistry, temperature range) may affect the maximum standing crop and activity of aquatic hyphomycetes on a given type of leaf litter as well as the shredder taxa that are present in a habitat. Consequently, these interactions are complex and a range of outcomes are possible. Unfortunately, in some studies, insufficient data concerning various aspects of the interaction (i.e. fungal growth and modification of leaves, palatability or food quality of fungal-leaf combination, nutrient concentrations in water used for conditioning, etc.) are provided. This often makes it difficult to interpret results.

6.2 Effects of Fungi on Shredders

Most studies which have examined fungal-shredder interactions have been concerned with the effects of fungal conditioning of leaves on the feeding ecology of shredders. Fungi have been shown to affect both the preferences and growth of shredders. Fungal growth on leaves can alter the palatability of leaf detritus, affecting shredder food choice. Fungi can also alter the food quality of leaf detritus, affecting shredder growth rates. Recent reviews of studies on fungal-shredder interactions have examined the role of fungi in the nutritional ecology of shredders (Bärlocher and Kendrick 1981; Bärlocher 1985), the nutritional ecology of aquatic detritivores (Cummins and Klug 1979; Anderson and Cargill 1987) and the role of invertebrates in detritus processing (Anderson and Sedell 1979).

6.2.1 Shredder Preferences

The palatability of leaf detritus has been shown to be affected by a number of factors including the species of microorganism colonizing leaves, conditioning time, the species of leaf and the shredder species consuming the detritus. Fungi play a greater role than bacteria in affecting food choice of both trichopteran and amphipod shredders (Mackay and Kalff 1973; Kostalos and Seymour 1976). These shredders also prefer the mycelium of certain fungal species over leaves (Bärlocher and Kendrick 1973a; Arsuffi and Suberkropp 1988). However, larvae of *Tipula* spp. behave differently. They do not consume leaves colonized by cultures of fungi (Findlay et al. 1986b; Arsuffi and Suberkropp 1989), but prefer leaves colonized by both fungi and bacteria (Lawson et al. 1984).

Fungal species, whether offered as mycelia or grown on leaves, are not equally palatable to the amphipod, *Gammarus pseudolimnaeus* (Bärlocher and Kendrick 1973a). In laboratory studies, the palatability of leaf detritus was found to be affected by fungal species, leaf species and environmental conditions (nutrients and temperature) during colonization. Of these factors, fungal species appeared to have a major influence on palatability since amphipod preferences for different leaf species could be changed by colonizing leaves with appropriate fungi. Under conditions in which fungi grew on leaves, the most palatable species were not always those that caused the greatest changes in leaves.

Trichopteran shredders can also distinguish among and select certain fungal species when offered a choice of leaves colonized by cultures of different fungi. In a comparison of ten fungal species, Suberkropp et al. (1983) found that all fungi grew well on aspen leaves, but differed greatly in palatability. Consequently, the palatability of the fungi could not be predicted from changes they caused, i.e. the amount of fungal biomass associated with leaves, softening of the leaf matrix, increases in N content or types and activities of fungal enzymes. When eight of these fungi were grown on oak leaves, two species, *T. marchalianum* and *Heliscus lugdunensis*, did not grow as well as the others and their palatability was also lower (Butler and Suberkropp 1986). This suggests that low capabilities of fungi to grow on or condition leaves results in low palatability. However, leaves may also remain unpalatable when they are colonized by fungi which exhibit high levels of growth and degradation. For example, oak and aspen leaves remained unpalatable for trichopteran larvae when they were colonized by certain species, e.g. *Lemonniera aquatica*, which grew well on both types of leaves (Suberkropp et al. 1983; Butler and Suberkropp 1986). Likewise, ash and maple leaves remained unpalatable to *G. pseudolimnaeus* when they were colonized by *T. marchalianum* which caused larger increases in the protein content of these leaves than other, more palatable, fungi (Bärlocher and Kendrick 1973a).

Another factor which affects the palatability of leaves colonized by fungi is the length of the conditioning period. In laboratory studies, the palatability of aspen leaves increased as fungi grew and modified leaves (Arsuffi and Suberkropp 1984). Once palatability increases, it may remain high for varying periods of time depending on the fungal species (Suberkropp and Arsuffi 1984). Since fungi grow at different rates, and their growth rates are affected by a number of factors including leaf species, temperature, and nutrient composition of the water in which leaves are suspended, changes in these factors would be expected to affect the time required for leaves to reach and remain at maximum palatability. Consequently, failure to take into account the time required for each fungus to colonize leaves under different conditions of incubation may lead to conflicting interpretations of the palatability of different fungi (Arsuffi and Suberkropp 1984).

Shredder taxa differ with respect to food choice. For example, larvae of *Tipula* spp. do not choose leaves colonized only by fungi whereas other shredders do. Not enough work on this aspect has been done to determine at what taxonomic level such differences exist. For trichopteran shredders, preferences of representatives of five genera from three families have been found to be similar (Arsuffi and Suberkropp 1984; Arsuffi and Suberkropp 1989; Suberkropp et al. 1983). However, preferences of trichopteran shredders were found to differ from those of an amphipod and a plecopteran shredder when all were offered the same choices of fungal species colonizing aspen leaves (Arsuffi and Suberkropp 1989). Such differences may represent a way in which shredders can partition food resources. Given the diets available, amphipod shredders were more selective and had narrower diet breadths than the other shredders examined in this study (Arsuffi and Suberkropp 1989). The amphipods consumed most of their preferred diet before switching to leaves colonized by other fungi whereas the insect shredders consumed some of all choices offered. More work is necessary to determine how significant such differences among shredder taxa can be. In comparing results from studies of different shredder and fungal taxa, phylogenetic differences must be considered before generalizations can be made.

Currently, we do not have an understanding of the mechanisms responsible for shredder preferences. For leaves to increase in palatability, they must first become conditioned. Partial hydrolysis of leaf polymers increases palatability (Bärlocher and Kendrick 1975b), but fungal biomass appears to be a greater determinant of palatability than modified leaf tissue (Arsuffi and Suberkropp 1988). Even though fungal biomass associated with leaves plays a key role in determining palatability, the amount does not appear to be the sole cue used by shredders in selecting detritus. Fungi which are highly palatable to a shredder may not grow on leaves as much as less palatable fungi (Bärlocher and Kendrick 1973a; Suberkropp et al. 1983). This, together with different preferences exhibited by shredder taxa when offered the same choices (Arsuffi and Suberkropp 1989), suggests that additional cues are used by shredders in selecting leaf detritus as food. Such cues appear to be species-specific for both fungi and shredders and have been proposed to be due to compounds produced by fungi that act as feeding stimulants or deterrents (Arsuffi and Suberkropp 1989; Bärlocher 1985). Cargill et al. (1985a) demonstrated that lipids, particularly unsaturated 18- and 20-carbon fatty acids, either as free acids or in triglycerides, act as feeding stimulants for the last instar of several trichopteran shredders. However, lipids did not stimulate feeding of early instars of these shredders nor of larvae of *T. abdominalis*. Neutral lipid extracts from five of the seven fungal species examined stimulated feeding in trichopteran shredders. Neutral lipid extracts from the two fungi which failed to stimulate feeding contained lower levels of triglycerides than extracts from the other fungi. However, direct comparisons of the palatability of lipid extracts from the different fungi were not made, so it is not known if this is a mechanism by which trichopteran shredders can distinguish among fungal species. To date, no other fungal metabolites have been shown to affect or modify the feeding behavior of shredders.

Most studies of preferences by shredders for fungi growing on leaves involve colonizing entire leaves or leaf disks with cultures of one fungus. In streams, fungal colonization of leaves results in a mosaic of patches colonized by different fungi (Shearer and Lane 1983; Chamier et al. 1984). Trichopteran shredders have the foraging capabilities to detect patches within a leaf colonized by several fungi (Arsuffi and Suberkropp 1985). Preferences exhibited for fungi occurring in patches were similar to preferences for fungi colonizing entire leaves. It is not known whether other shredders have this capability, but the partial skeletonization of leaves in streams suggests that shredders can forage selectively for palatable patches within a leaf.

6.2.2 Shredder Nutrition

The food quality of leaf detritus has been assessed by various methods. Cargill et al. (1985b) emphasized the importance of determining whether the shredder can complete development and used fecundity to compare the food quality of different diets. Most studies, however, have evaluated food quality over shorter periods of time, measuring shredder growth rates or survivorship during a portion of the total development of the animal. Like palatability, the food quality of leaf detritus determined by the latter methods has been shown to be affected by a number of factors including fungal species composition, leaf species and conditioning time.

6.2.2.1 Growth and Survivorship

Shredder growth and survivorship has been found to vary greatly depending on the fungal species in the diet (Bärlocher and Kendrick 1973b; Rossi and Fano 1979; Arsuffi and Suberkropp 1986). Such results suggest that fungi differ markedly in their food quality to amphipod, isopod and trichopteran shredders. For example, Bärlocher and Kendrick (1973b) found that four out of ten fungi supported significantly better growth of *G. pseudolimnaeus* than unconditioned elm or maple leaves. Growth of the amphipod on an additional four fungal species was not significantly different from growth on leaves, and two species were unsuitable as food since animals fed these fungi died as rapidly as animals which were not fed. Two trichopteran shredders exhibited a range of growth rates (1.9–7.7% day^{-1}) when they were fed diets of aspen leaves colonized by different fungi (Arsuffi and Suberkropp 1986). For *Psychoglypha* sp. larvae, mortality ranged from 0–100% depending on the fungal species in the diet.

For amphipod and trichopteran shredders, some agreement has been found between preference rankings (Bärlocher and Kendrick 1973a; Suberkropp et al. 1983) and food quality (Bärlocher and Kendrick 1973b; Arsuffi and Suberkropp 1986). Generally, fungi which are most preferred by a particular shredder support the highest growth rates in that shredder. However, shredders exhibit a range of growth rates on less palatable fungi, and these growth rates do not appear to be correlated with the palatability of the fungus as determined in preference trials.

Two species of isopod shredders (*Asellus aquaticus* and *Proasellus coxalis*) exhibited differential growth and survivorship depending on the fungal species they were fed, leading Rossi and Fano (1979) to suggest that these isopods could partition food resources in habitats where populations of both species coexisted. In further studies, allopatric populations of these isopod species exhibited fewer differences in performance on diets consisting of mycelia of different fungal species than did sympatric populations (Rossi et al. 1983). Thus, it appears that populations of the same shredder species can exhibit different trophic specialization depending on its habitat and potential competitors. Such adaptation could result from the phenotypic variation in performance on different fungal diets found within populations of isopod shredders (Rossi 1985). Whether other shredders exhibit such variation in their abilities to assimilate different fungal diets has not been determined.

In other studies, shredder performance on diets of fungal mycelium has been found to be better (Cargill et al. 1985b; Arsuffi and Suberkropp 1988), similar (Kostalos and Seymour 1976) or worse (Willoughby and Sutcliffe 1976; Marcus and Willoughby 1978; Sutcliffe et al. 1981) than their performance on diets of leaves conditioned in different ways. For example, *Gammarus pulex* was found to grow better on decaying oak and elm leaves from a stream than on mycelia of *Clavariopsis aquatica*, *Tricladium giganteum* (Willoughby and Sutcliffe 1976) or *Lemonniera aquatica* (Sutcliffe et al. 1981). Likewise the isopod, *Asellus aquaticus*, was found to grow better on oak leaves conditioned in a lake and on mycelia of the Oomycete, *Saprolegnia* sp. than on mycelia of *L. aquatica* (Marcus and Willoughby 1978). Since shredders exhibit a range of growth rates when different fungi are compared (see above), such results may reflect comparisons among fungal species of different food quality. For example, *L. aquatica* is a poor and unpalatable food source for some shredders as it supported the lowest growth rate in trichopteran shredders of the seven fungi compared as diets (Arsuffi and Suberkropp 1986). It was also low in the preference rankings of amphipod, plecopteran and trichopteran shredders offered choices of leaves colonized by one of eight fungi (Arsuffi and Suberkropp 1989). In contrast, *C. aquatica* appears to be a palatable and relatively good food source for amphipod shredders. *Gammarus pseudolimnaeus* exhibited higher growth rates on diets of *C. aquatica* mycelia than on unconditioned elm and maple leaves (Bärlocher and Kendrick 1973b). Aspen leaves colonized by this fungus were the second highest in preference rankings of amphipods (Arsuffi and Suberkropp 1989). It is possible that the decaying leaves which Willoughby and Sutcliffe (1976) compared with *C. aquatica* were fully conditioned by fungi with greater food quality than *C. aquatica*. The fungal species composition and the stage of conditioning of the decaying leaves

Table 6.2. Digestive capabilities of shredders

Shredder	Gut pH	Protein	Amylose	Laminarin	Xylan	Pectin	CMC	Crystalline Cellulose	References[a]
Gammarus[b]	6.8–7.2	++	++	++	+	–	+	+	[1, 2]
Tipula mid	7.2–11.2	+++	++	++	–	–	±	–	[1, 3, 6]
hind	6.5–7.5								
Pteronarcys	6.4–7.7	+++	++	++	+	+	+	–	[4, 6]
Pycnopsyche	6.1–7.8	+++	++	++	+	–	±	–	[5, 6]

[a] [1] Bärlocher and Porter (1986); [2] Chamier (1991); [3] Martin et al. (1980); [4] Martin et al. (1981a); [5] Martin et al. (1981b); [6] Sinsabaugh et al. (1985).
[b] Results have been summarized for at least two species of each shredder.

used in their study were not reported. It is interesting that the growth rates of *G. pulex* (Willoughby and Sutcliffe 1976) and *G. pseudolimnaeus* (Bärlocher and Kendrick 1973b) were similar when both were fed mycelia of *C. aquatica*, even though the two studies reached contrasting conclusions concerning the quality of fungi as food for *Gammarus*. These comparisons emphasize the complex nature of fungal-shredder interactions and indicate the variety of factors, particularly with respect to conditioning of leaves, that can make results difficult to interpret.

6.2.2.2 Digestive Adaptations

Another approach which has been used to examine the potential role of fungi in the feeding ecology of shredders has been to determine the capabilities of shredders for digesting components of leaf detritus. The digestive tracts of all shredders which have been examined contain relatively high activities of proteinases (Table 6.2). In most of these animals, the pH of the gut has been found to be neutral to slightly alkaline. Only larvae of *Tipula* spp. have been found to have highly alkaline pH (up to pH 11.2) in the midgut (Martin et al. 1980; Sinsabaugh et al. 1985; Bärlocher and Porter 1986). Such alkaline conditions should allow greater release of proteins from complexes with phenolic compounds and be more favorable for the proteinases with alkaline pH optima which are found in these animals (Martin et al. 1980). Bärlocher et al. (1989) found that gut extracts from *Tipula caloptera*, but not *Gammarus tigrinus*, caused significant decreases in the protein content of complexes formed between leaf leachates and a fungal cellulase. These observations suggest that *Tipula* larvae are adapted for digestion of leaf proteins whereas other shredders with less alkaline conditions in their digestive tract can only digest microbial proteins which have not been complexed with phenolics (Martin et al. 1980, 1981b).

With respect to polysaccharides (Table 6.2), the digestive fluid of all shredders which have been examined possesses moderate activity to hydrolyze amylose (an α-1,4-glucan) and laminarin (a β-1,3-glucan), polymers which are more common in fungi and algae than in dead leaves (Martin et al. 1981a,b). The digestive fluid of trichopteran, plecopteran and amphipod shredders generally possess weak activity for the hydrolysis of polysaccharides common in leaves (e.g. cellulose, xylan, pectin). Cellulase activities in the guts of trichopteran and plecopteran shredders have been suggested to be acquired by these shredders from the consumption of detritus containing microbial enzymes (Sinsabaugh et al. 1985). Fungal carbohydrases have been demonstrated to retain activity in the guts of *Gammarus fossarum* (Bärlocher 1982b) suggesting that amphipods may also acquire microbial enzymes during ingestion of fully conditioned detritus. Amphipods possess a wider array of degradative enzymes in their digestive tract than insect shredders (Bärlocher and Porter 1986) and have the capability to produce cellulase complexes that are required for degradation of crystalline cellulose (Monk 1977; Chamier 1991). Gut extracts of *Tipula* generally have not been found to contain detectable levels of enzymes capable of degrading cellulose, xylan or pectin (Martin et al. 1980; Bärlocher and Porter 1986). However, using radiolabelled cellulose, Sinsabaugh et al. (1985) measured assimilation of cellulose by *T. abdominalis*, and suggested that such activity may be associated with the dense microbiota indigenous to the hindgut fermentation chamber in this animal (Klug and Kotarski 1980). This was supported by finding that the majority of label derived from cellulose in ligated guts of *T. abdominalis* was converted to organic or amino acids (Sinsabaugh et al. 1985). In ligated guts of *Pteronarcys proteus*, which contain low densities of gut bacteria, a significant fraction of the label released from cellulose was detected in sugars. Overall, these results suggest that with the exception of the digestion of leaf proteins by *Tipula* and cellulose by *Gammarus*, leaf-shredding invertebrates are poorly equipped to digest leaves without microbial conditioning.

6.2.2.3 Contribution of Fungi to Shredder Nutrition

Although some fungal species are better food sources than dead leaves, a number of authors have suggested that fungal biomass is not present in sufficient concentrations on leaf detritus to account fully for shredder assimilation (Bärlocher and Kendrick 1981; Lawson et al. 1984; Findlay et al. 1986a,b). Currently, there are few estimates of the amount of fungal biomass associated with leaves decomposing in streams, particularly in studies where shredder growth rates or assimilation have been determined. In addition, the maximum levels of microbial biomass observed on leaf litter can vary depending on the chemical characteristics (pH, alkalinity, N and P concentrations) of the stream and leaf species (Rosset et al. 1982; Suberkropp 1991). Estimates of fungal biomass based on microscopic determination of hyphal biovolume after clearing or grinding leaves range from much less than 1% to up to 5% of the detrital mass (Iversen 1973; Bärlocher and Kendrick 1981; Findlay and Arsuffi 1989). Fungal biomass has been determined to account for 90–95% of total microbial biomass during early stages of colonization (Findlay and Arsuffi 1989). If ATP contents of leaves colonized in streams (Suberkropp and Klug 1976; Rosset et al. 1982; Lawson et al. 1984) are converted entirely to fungal biomass, concentrations of fungal biomass can be as high as 10–12% of detrital mass. Likewise, measurements of ergosterol concentrations associated with leaves decomposing in streams suggest that fungal biomass can reach 10% of detrital mass (Gessner and Chauvet 1990). Chemical methods for estimating fungal biomass associated with leaves obtain average values for all components of leaf tissue (e.g. vascular, mesophyll). Since some shredders can forage selectively (Arsuffi and Suberkropp 1985), it is possible that shredders may actually consume pieces of leaf detritus that contain higher concentrations of fungal biomass than 10%.

There is no doubt that leaves can be used as a food source by some shredders. Unconditioned leaves have been shown to be assimilated by *Pteronarcys proteus* (Golladay et al. 1983) and two species of *Gammarus* (Bärlocher and Kendrick 1973b; Willoughby and Sutcliffe 1976; Sutcliffe et al. 1981). However, two isopod species died with sterile leaves as their sole diet (Rossi and Fano 1979). Similar types of studies have not been carried out with other shredders, but *T. abdominalis* did not grow on a diet consisting of hickory leaves which had been conditioned for only 1 week in a stream (Lawson et al. 1984).

Fungal mycelia are generally considered to be assimilated with greater efficiency than unconditioned leaves. Bärlocher and Kendrick (1975a) estimated that assimilation efficiencies (based on ash-free dry mass) of *G. pseudolimnaeus* on diets of fungal mycelia were 65–77% whereas those on diets of unconditioned elm and maple leaves were 13–14%. Based on these values and estimates that fungal biomass could account for up to 10% of leaf detritus, Bärlocher and Kendrick (1981) calculated that microbial cells could only account for a maximum of 25% of the nutrients obtained by shredders feeding on conditioned leaf detritus. However, caution should be used in making calculations with assimilation efficiencies in this manner, since there is evidence that shredders can adjust their ingestion rate and, consequently, their assimilation rate depending on the conditioning stage of the detritus (Anderson and Grafius 1975; Golladay et al. 1983; Lawson et al. 1984; Arsuffi and Suberkropp 1986). Golladay et al. (1983) concluded that assimilation efficiency was a poor indication of food quality since the plecopteran shredder they examined did not change assimilation efficiency, but did alter ingestion and assimilation rates depending on the conditioning stage of the leaf detritus. Therefore, even though shredders may be able to obtain nutrition from unconditioned leaves, the more rapid ingestion (and egestion) rates exhibited for conditioned detritus may preclude the same degree of digestion and assimilation of leaf material when leaf-detritus diets contain high concentrations of fungal biomass. Further studies are necessary to examine this possibility more fully.

Findlay et al. (1986a,b) attempted to measure the contribution of fungal biomass to the assimilation of two shredder taxa directly by labeling fungi growing on leaf disks with ^{14}C-labelled glucose. They concluded that fungal biomass associated with leaves colonized in the laboratory could not account for most of the respiratory requirements of the animals (Table 6.3). However, the procedures used by Findlay et al. (1986a,b) to grow fungi on leaves resulted in low amounts of fungal biomass (Table 6.3). The greatest amount of fungal biomass detected occurred with *Mucor* sp. grown on oak leaves, even though *Mucor* spp. are generally considered to lack capabilities for degradation of the structural components of leaves. For this fungus, biomass (as fungal C per leaf disk) was determined from ATP concentrations, whereas for other fungi, biomass was determined from hyphal biovolume using microscopic techniques. Growth of *Mucor* sp. on oak leaves resulted

Table 6.3. Fungal biomass and respiratory carbon requirements of shredders provided by fungi grown on leaves in the laboratory. Data from Findlay et al. (1986a,b)

Shredder	Fungus	Leaf	Fungal biomass µg C disk^{-1}	Respiratory C provided by fungus (%)
Lirceus sp.	*Mucor* sp.	Oak	245	57
		Willow	48	16
Lirceus sp.	*Triscelophorus* sp.	Oak	7	1
		Willow	2	1
Peltoperla sp.	*Articulospora tetracladia*	Oak	14	27–39
		Birch	25	25–36

in a maximum of 72 µg ATP g^{-1} (=245 µg C disc^{-1}, Table 6.3). Most studies have reported maximum ATP concentrations associated with leaves colonized in streams to range from 65–175 µg ATP g^{-1}, (Suberkropp and Klug 1976; Rosset et al. 1982; Lawson et al. 1984; Suberkropp 1991), although I have found the maximum ATP concentration of leaves in one stream with very low nutrients to be 16 µg g^{-1} (Suberkropp 1991). When fungi are grown on leaves in the laboratory, concentrations of 200–300 µg ATP g^{-1} can be reached (Suberkropp et al. 1983). Therefore, the diet of *Mucor* sp. grown on oak leaves contained fungal biomass that was at the low end of the range of values for fully conditioned leaves. When fed this diet, an isopod shredder assimilated fungal carbon that was calculated to account for 57% of its respiratory requirements. The other diets examined by Findlay et al. (1986a,b), particularly those containing aquatic hyphomycetes, contained even lower concentrations of fungal biomass (0.01–0.1x, see Table 6.3). In view of these low concentrations, it is somewhat surprising that fungal biomass could account for as large a percentage of the respiratory requirements of these animals as was estimated, especially for the diets containing *Articulospora tetracladia*. Findlay et al. (1986a,b) did not report whether the shredders they examined gained or lost weight on any of the diets nor whether the fungi were of high or low food quality for these shredders.

Lawson et al. (1984) used ^{15}N to label N in the microbial component of leaves in culture and estimated the contribution of microbial biomass to the assimilation of N by *T. abdominalis*. Diets of leaves colonized by both fungi and bacteria in the laboratory were higher quality food than leaves conditioned in a stream as they produced higher relative growth rates in *T. abdominalis* than stream-conditioned leaves. Based on partitioning of labeled N, Lawson et al. (1984) concluded that microbial biomass associated with leaves colonized in the laboratory accounted for 11–27% of larval growth. This calculation was based on the assumption that assimilation of all components of microbial biomass was the same as the assimilation of microbial N (Lawson et al. 1984). The validity of this assumption, however, should be questioned since *Tipula* possesses digestive adaptations (high pH in midgut and alkaline proteinases, see above) that enhance the release and hydrolysis of proteins from complexes with phenolics. This suggests that *Tipula* larvae can obtain a higher percentage of their N from leaf proteins than other shredders and that their capabilities for assimilating N may differ from their assimilation of nonproteinaceous components. The role of the dense microbial populations in the hindgut fermentation chamber of *Tipula* has not been fully determined, but they appear to produce fermentation products that can be used by the animal (Lawson and Klug 1989). Such gut symbionts may allow *Tipula* larvae to rely less on microbial conditioning of detritus before ingestion than other shredders with less extensive microbiota in their digestive tracts.

If fungal biomass is not present in sufficient quantities to account fully for shredder assimilation, then fungal conditioning of leaves must improve the food quality of leaf detritus by other ways. Two mechanisms have been proposed (Bärlocher 1985): fungal degradation of leaf polymers into more digestible forms and continued activity of extracellular enzymes in the digestive tracts of shredders. Data supporting both possibilities have been obtained (Bärlocher and Kendrick 1975b; Bärlocher 1982b; Sinsabaugh et al. 1985). Together with the observations that some shredders can assimilate leaves (see above) these data suggest that shredders can obtain nutrition from both the decomposing leaf and the associated microorganisms, particularly fungi. The relative contribution of each to the total assimilation by shredders may vary and is surely affected by stage of conditioning, species of fungi growing on leaves and digestive adaptations of the shredder. However, before the quantitative significance of fungi in the diet of shredders is discounted (Findlay et al. 1986b), the range of these contributions should be determined under conditions that are similar to those encountered in streams and in which the shredders have been demonstrated to grow.

6.3 Effects of Shredders on Fungi

Very little is known concerning the effects of shredder feeding on fungi. Bärlocher (1982a) examined the fungal communities occurring on three types of litter that were exposed to shredder feeding in coarse mesh bags or were protected from shredder feeding in fine mesh bags. These studies were carried out in two hardwater streams with high densities of amphipod shredders and two softwater streams with somewhat lower densities of shredders. In all streams, the fungal communities on oak and larch litter exhibited lower mass-specific rates of sporulation in coarse mesh bags than those in fine mesh bags. Both types of litter were also broken down more rapidly in coarse mesh bags so that total fungal reproduction (rate of sporulation $\times$ time) was greatly reduced in the coarse mesh treatments. Shredders did not feed on spruce needles and similar rates of sporulation occurred in both coarse and fine mesh treatments.

Bärlocher (1980) found that the species richness of aquatic hyphomycetes on oak and larch litter was lower in coarse mesh bags than in fine mesh bags. Species which did not occur on litter in the coarse mesh treatments were those which typically occurred late and were rare. These observations led Bärlocher to conclude that shredder feeding had two effects on the fungal communities: (1) they prey on early common species reducing their overall sporulation and (2) they compete with late rare species by consuming the leaf substratum so that it is not available for subsequent colonization.

Wallace et al. (1982) examined the effects of shredders on leaf litter breakdown by eliminating aquatic insects in one stream with the insecticide methoxychlor. In the absence of shredders, breakdown rates of leaf detritus were lower in the treated stream than in a control stream which received no insecticide. In a subsequent study in which an additional stream was treated with methoxychlor, concentrations of aquatic hyphomycete conidia in the water were found to be higher than concentrations in untreated streams (Wallace and Suberkropp, unpublished data). These results provide further evidence that shredder feeding reduces overall reproduction by aquatic hyphomycetes.

Although shredder feeding in the laboratory has been shown to be selective for certain fungal species, it is not clear whether selective feeding by shredders occurs in streams or what its effects might be. In the study reported by Bärlocher (1982a), changes in community composition of the dominant aquatic hyphomycetes on oak or larch litter were presented for only one stream. During the period of time when shredder feeding was most intense, comparisons between data from coarse mesh and fine mesh treatments indicate an apparent reduction in *Alatospora acuminata* as a percentage of the total. Whether these changes were statistically significant and due to selective feeding cannot be determined, but these data are at least suggestive of such an effect.

6.4 Conclusions

Interactions between aquatic hyphomycetes and shredders are dynamic and complex. They have an additional level of complexity over conventional predator-prey interactions in that the leaf on which fungi are growing serves as a resource for both the fungus and the shredder. Changes in leaves as a result of fungal activity also affect their use as a resource for shredders. However, all fungi do not transform leaf litter into more palatable and nutritious foods. Improvements in food quality of leaves as a result of fungal growth and degradation of leaf polymers may be balanced by the production of fungal metabolites which deter feeding (antifeedants, toxins). Since there is sufficient evidence to demonstrate that fungi vary considerably in both palatability and food quality, fungal species composition is important in determining the outcome of these interactions. In addition, shredder taxa differ in their selectivity, mechanism of food choice and in their ability to assimilate different diets. Shredder feeding on leaf detritus is further affected by conditioning stage, characteristics of the leaf and environmental factors during conditioning. Better understanding of fungal-shredder interactions requires that investigators define and control these variables. Many of the discrepancies currently present in the literature appear to be due to differences in choice of fungal species and methods used to condition leaves for a particular shredder.

Digestive adaptations in shredders suggest that most are poorly equipped to digest unconditioned leaf material. If this is the case, then fungal activities, i.e. growth and enzymatic transformation of leaves into more digestible foods, are necessary. The evidence that is currently available suggests that, when feeding on leaf detritus, shredders obtain more of their nutrition from leaf tissue than from fungal biomass. However, this conclusion is based on a limited number of studies with few shredder species. Estimates of the relative contributions of fungal biomass, unmodified leaf tissue and enzymatically modified leaf tissue to shredder assimilation under a range of environmental conditions are needed to more fully understand the role of fungi in the feeding ecology of shredders.

Shredder feeding has been shown to reduce the reproductive output of fungi in several streams. Selective feeding by shredders also has the potential to alter fungal community composition. Whether this occurs in streams remains unknown. Factors such as shredder densities, intensity of feeding and amount of leaf litter in a stream would be expected to affect the impact of shredder feeding on fungal populations. How these interactions affect fungal community structure requires additional research if we are to increase our understanding of the ecology of aquatic hyphomycetes in streams.

Acknowledgements. I thank Drs. Eric Chauvet and Felix Bärlocher for comments on the manuscript and the National Science Foundation (Grants BSR 8307325 and BSR 8705486) for support of my research on fungal-shredder interactions.

References

Anderson NH, Cargill AS (1987) Nutritional ecology of aquatic detritivorous insects. In: Slansky F Jr, Rodriguez JG (eds) Nutritional ecology of insects, mites and spiders. Wiley and Sons, New York, pp 903–925.

Anderson NH, Grafius E (1975) Utilization and processing of allochthonous material by stream Trichoptera. Verh Int Ver Limnol 19:3083–3088

Anderson NH, Sedell JR (1979) Detritus processing by macroinvertebrates in stream ecosystems. Annu Rev Entomol 24:351–377

Arsuffi TL, Suberkropp K (1984) Leaf processing capabilities of aquatic hyphomycetes: interspecific differences and influence on shredder feeding preferences. Oikos 42:144–154

Arsuffi TL, Suberkropp K (1985) Selective feeding by stream caddisfly (Trichoptera) detritivores on leaves with fungal-colonized patches. Oikos 45:50–58

Arsuffi TL, Suberkropp K (1986) Growth of two stream caddisflies (Trichoptera) on leaves colonized by different fungal species. J N Am Benthol Soc 5:297–305

Arsuffi TL, Suberkropp K (1988) Effects of fungal mycelia and enzymatically degraded leaves on feeding and performance of caddisfly (Trichoptera) larvae. J N Am Benthol Soc 7:205–211

Arsuffi TL, Suberkropp K (1989) Selective feeding by shredders on leaf-colonizing stream fungi: comparison of macroinvertebrate taxa. Oecologia 79:30–37

Bärlocher F (1980) Leaf-eating invertebrates as competitors of aquatic hyphomycetes. Oecologia 47:303–306

Bärlocher F (1982a) Conidium production from leaves and needles in four streams. Can J Bot 60:1487–1494

Bärlocher F (1982b) The contribution of fungal enzymes to the digestion of leaves by *Gammarus fossarum* Koch (Amphipoda). Oecologia 52:1–4

Bärlocher F (1985) The role of fungi in the nutrition of stream invertebrates. Bot J Linn Soc 91:83–94

Bärlocher F, Kendrick B (1973a) Fungi and food preferences of *Gammarus pseudolimnaeus*. Arch Hydrobiol 72:501–516

Bärlocher F, Kendrick B (1973b) Fungi in the diet of *Gammarus pseudolimnaeus* (Amphipoda). Oikos 24:295–300

Bärlocher F, Kendrick B (1975a) Assimilation efficiency of *Gammarus pseudolimnaeus* (Amphipoda) feeding on fungal mycelium or autumn-shed leaves. Oikos 26:55–59

Bärlocher F, Kendrick B (1975b) Leaf-conditioning by microorganisms. Oecologia 20:359–362

Bärlocher F, Kendrick B (1981) Role of aquatic hyphomycetes in the trophic structure of streams. In: Wicklow DT, Carroll GC (eds) The fungal community: its organization and role in the ecosystem. Marcel Dekker, New York, pp 743–760

Bärlocher F, Porter CW (1986) Digestive enzymes and feeding strategies of three stream invertebrates. J N Am Benthol Soc 5:58–66

Bärlocher F, Tibbo PG, Christie SH (1989) Formation of phenol-protein complexes and their use by two stream invertebrates. Hydrobiologia 173:243–249

Boling RH, Goodman ED, Van Sickle JA, Zimmer JO, Cummins KW, Petersen RC, Reice SR (1975) Toward a model of detritus processing in a woodland stream. Ecology 56:141–151

Butler SK, Suberkropp K (1986) Aquatic hyphomycetes on oak leaves: comparison of growth, degradation and palatability. Mycologia 78:922–928

Cargill AS II, Cummins KW, Hanson BJ, Lowry RR (1985a) The role of lipids as feeding stimulants for shredding aquatic insects. Freshwater Biol 15:455–464

Cargill AS II, Cummins KW, Hanson BJ, Lowry RR (1985b) The role of lipids, fungi, and temperature in the nutrition of a shredder caddisfly, *Clistoronia magnifica*. Freshwater Invert Biol 4:64–78

Chamier A-C (1991) Cellulose digestion and metabolism in the freshwater amphipod *Gammarus pseudolimnaeus* Bousfield. Freshwater Biol 25:33–40

Chamier A-C, Dixon PA, Archer SA (1984) The spatial distribution of fungi on decomposing alder leaves in a freshwater stream. Oecologia 64:92–103

Chauvet E (1987) Changes in the chemical composition of alder, poplar and willow leaves during decomposition in a river. Hydrobiologia 148:35–44

Cummins KW (1973) Trophic relations of aquatic insects. Annu Rev Entomol 18:183–206

Cummins KW, Klug MJ (1979) Feeding ecology of stream invertebrates. Annu Rev Ecol Syst 10:147–172

Findlay SEG, Arsuffi TL (1989) Microbial growth and detritus transformations during decomposition of leaf litter in a stream. Freshwater Biol 21:261–269

Findlay SEG, Meyer JL, Smith PJ (1986a) Contribution of fungal biomass to the diet of a freshwater isopod (*Lirceus* sp.). Freshwater Biol 16:377–385

Findlay SEG, Meyer JL, Smith PJ (1986b) Incorporation of microbial biomass by *Peltoperla* sp. (Plecoptera) and *Tipula* sp. (Diptera). J N Am Benthol Soc 5:306–310

Gessner MO, Chauvet E (1990) Biomass estimates of aquatic hyphomycetes associated with leaf-litter. 4th Int Mycol Congr Abstr, p 121

Golladay SW, Webster JR, Benfield EF (1983) Factors affecting food utilization by a leaf shredding aquatic insect: leaf species and conditioning time. Holarct Ecol 6:157–162

Iversen TM (1973) Decomposition of autumn-shed leaves in a springbrook and its significance for the fauna. Arch Hydrobiol 72:305–312

Kaushik NK, Hynes HBN (1971) The fate of the dead leaves that fall into streams. Arch Hydrobiol 68:465–515

Klug MJ, Kotarski S (1980) Bacteria associated with the gut tract of larval stages of the aquatic cranefly *Tipula abdominalis* (Diptera; Tipulidae). Appl Environ Microbiol 40:408–416

Kostalos M, Seymour RL (1976) Role of microbial enriched detritus in the nutrition of *Gammarus minus* (Amphipoda). Oikos 27:512–516

Lawson DL, Klug MJ (1989) Microbial fermentation in the hindguts of two stream detritivores. J N Am Benthol Soc 8:85–91

Lawson DL, Klug MJ, Merritt RW (1984) The influence of the physical, chemical, and microbiological characteristics of decomposing leaves on the growth of the detritivore *Tipula abdominalis* (Diptera: Tipulidae). Can J Zool 62:2339–2343

Mackay RJ, Kalff J (1973) Ecology of two related species of caddis fly larvae in the organic substrates of a woodland stream. Ecology 54:499–511

Marcus JH, Willoughby LG (1978) Fungi as food for the aquatic invertebrate *Asellus aquaticus*. Trans Br Mycol Soc 70:143–146

Martin MM, Martin JS, Kukor JJ, Merritt RW (1980) The digestion of protein and carbohydrate by the stream detritivore, *Tipula abdominalis* (Diptera, Tipulidae). Oecologia 46:360–364

Martin MM, Kukor JJ, Martin JS, Lawson DL, Merritt RW (1981a) Digestive enzymes of larvae of three species of caddisflies (Trichoptera). Insect Biochem 11:501–505

Martin MM, Martin JS, Kukor JJ, Merritt RW (1981b) The digestive enzymes of detritus-feeding stonefly nymphs (Plecoptera; Pteronarcyidae). Can J Zool 59:1947–1951

Meyer JL (1980) Dynamics of phosphorus and organic matter during leaf decomposition in a forest stream. Oikos 34:44–53

Monk DC (1977) The digestion of cellulose and other dietary components, and pH of the gut in the amphipod *Gammarus pulex* (L.). Freshwater Biol 7:431–440

Rosset J, Bärlocher F, Oertli JJ (1982) Decomposition of conifer needles and deciduous leaves in two Black Forest and two Swiss Jura streams. Int Rev Gesamten Hydrobiol 67:695–711

Rossi L (1985) Interactions between invertebrates and microfungi in freshwater ecosystems. Oikos 44:175–184

Rossi L, Fano AE (1979) Role of fungi in the trophic niche of the congeneric detritivorous *Asellus aquaticus* and *A. coxalis* (Isopoda). Oikos 32:380–385

Rossi L, Basset A, Nobile L (1983) A coadapted trophic niche in two species of Crustacea (Isopoda): *Asellus aquaticus* (L.) and *Proasellus coxalis* Dolff. Evolution 37:810–820

Shearer CA, Lane LC (1983) Comparison of three techniques for the study of aquatic hyphomycete communities. Mycologia 75:498–508

Sinsabaugh RL, Linkins AE, Benfield EF (1985) Cellulose digestion and assimilation by three leaf-shredding aquatic insects. Ecology 66:1464–1471

Suberkropp K (1991) Relationships between growth and sporulation of aquatic hyphomycetes on decomposing leaf litter. Mycol Res 95:843–850

Suberkropp K, Arsuffi TL (1984) Degradation, growth and changes in palatability of leaves colonized by six aquatic hyphomycete species. Mycologia 76:398–407

Suberkropp K, Klug MJ (1976) Fungi and bacteria associated with leaves during processing in a woodland stream. Ecology 57:707–719

Suberkropp K, Klug MJ (1981) Degradation of leaf litter by aquatic hyphomycetes. In: Wicklow DT, Carroll GC (eds) The fungal community: its organization and role in the ecosystem. Marcel Dekker, New York, pp 761–776

Suberkropp K, Godshalk GL, Klug MJ (1976) Changes in the chemical composition of leaves during processing in a woodland stream. Ecology 57:720–727

Suberkropp K, Arsuffi TL, Anderson JP (1983) Comparison of degradative ability, enzymatic activity, and changes in palatability of aquatic hyphomycetes grown on leaf litter. Appl Environ Microbiol 46:237–244

Sutcliffe DW, Carrick TR, Willoughby LG (1981) Effects of diet, body size, age and temperature on growth rates in the amphipod *Gammarus pulex*. Freshwater Biol 11:183–214

Triska FJ (1970) Seasonal distribution of aquatic hyphomycetes in relation to the disappearance of leaf litter from a woodland stream. PhD Thesis, University of Pittsburgh, Pittsburgh

Wallace JB, Webster JR, Cuffney TF (1982) Stream detritus dynamics: regulation by invertebrate consumers. Oecologia 53:197–200

Willoughby LC, Sutcliffe DW (1976) Experiments on feeding and growth of the amphipod *Gammarus pulex* (L.) related to its distribution in the River Duddon. Freshwater Biol 6:577–586

7 Attachment and Germination of Conidia

S.J. Read, S.T. Moss, and E.B.G. Jones

7.1 Introduction

Aquatic hyphomycetes have been shown to be successful colonizers of leaf material in freshwater streams (Ingold 1956, 1966, 1975, 1984; Shearer and Webster 1985a,b; Webster 1981). The initial requirements for colonization of a fresh substratum are the formation and liberation of conidia by the parent colony; these reach a peak concomitant with leaf fall (Chamier et al. 1984; Shearer and Webster 1985a,b; Chap. 3). Increases in amounts and variety of substrata, water turbulence and rapid growth rates have been implicated in this seasonal periodicity of increased inoculum (Sanders and Webster 1980; Webster and Descals 1981; Suberkropp 1984). It has also been suggested (Shearer and Webster 1985b; Chaps. 4, 5) that species of aquatic hyphomycetes may possess a resting stage such as the teleomorphic and chlamydosporic stage for their survival during unfavourable conditions and to provide an inoculum when new substrata become available.

Ingold (1953, 1954, 1976), Webster (1959, 1987) and Webster and Descals (1981) suggested several factors which contribute to the success of aquatic hyphomycetes. These included the influence of conidium shape on attachment to substrata, rapid colonization and sporulation once substrata become available, ability to grow and sporulate at temperatures down to 0 °C and an effective method of transmission between unconnected water courses, possibly by means of mycelium embedded in leaf tissue or on the feet of waterfowl. The production of a wide range of extracellular enzymes (Zemek et al. 1985; Chaps. 1, 4, 6, 8) is another factor which contributes to the success of the aquatic hyphomycetes.

Based on a study of conidial germination and attachment, four additional factors which contribute to aquatic hyphomycete success were described (Read 1990). These factors were: the ability of released conidia to effect an early, secure attachment following contact with a substratum; rapid germination rate of conidia coupled with high percentage germination; mucilage formation which aids attachment of spores and germ tubes; appressorium formation, which increases adhesion to a substratum. These four factors will be discussed in greater detail in this chapter.

7.2 Conidial Attachment

The most obvious adaptation of aquatic hyphomycetes to their habitat is the size and morphology of their conidia (Webster 1987), many of which are either tetraradiate (Fig. 7.1, *1–4*), branched (Fig. 7.2, *12*) or filiform (Fig. 7.1, *6*) and exceed 100 μm in length although small ovoid conidia also exist (Fig. 7.1, *5*, *7*, *8*; Fig. 7.2, *14*). Tetraradiate conidia are the most frequently observed spore forms and although their increased buoyancy is considered to have a negligible effect under turbulent conditions (Webster 1959) they are trapped efficiently by the air bubbles in surface foam (Iqbal and Webster 1973) which acts as a dispersive agent.

The second most common spore form typical of aquatic hyphomycetes is sigmoid, a configuration which also aids attachment (Webster and Davey 1984; Webster 1987). Parallel and/or convergent evolution has resulted in the formation of long sigmoid/filiform propagules in both aquatic fungi and in higher plants which rely upon water for dispersal. This may be due to the direct relationship between increasing propagule length and the probability of impaction for aquatic sigmoid propagules (Cox 1983).

Subsequent to or concomitant with impaction the conidia must attach to the substratum. Webster (1959) showed that tetraradiate conidia were trapped on to submerged collodion-coated rods more efficiently than sigmoid and then ovoid forms. Quantitative data were obtained from attachment studies using a Fowler Radial Flow Chamber (Fowler and McKay 1980; Read et al. 1988). These data showed that tetraradiate conidia attached initially more strongly to the substratum than other conidial forms, the strength of attachment of the conidia then increased with time (see Fig. 7.5, *29*). The Mean Detachment Value, that is the shear stress required to detach 10% of the propagules, is a useful indicator of attachment for comparison of aquatic micro-organisms. The initial strength of attachment of the tetraradiate conidia of the freshwater hyphomycetes (Read 1990) was similar to those of the marine fungal spores examined by Hyde et al. (1989). It was speculated that the possession of multiple contact points increased the chances of early adhesion to a potential substratum (Read et al. 1988).

7.3 Germination

Although conidia of *Anguillospora crassa*, *Heliscus lugdunensis* and *Tumularia aquatica* show slow initial germination rates, generally conidia of aquatic hyphomycete species show unusually high percentage germination (see Fig. 7.5, *26*) and germination rates (Read 1990). Furthermore, the germination rates of aquatic hyphomycetes are not significantly reduced by low temperature. In terrestrial species germination rates decrease as the temperature drops, particularly below 15 °C (Byrne and Jones 1975; Magan

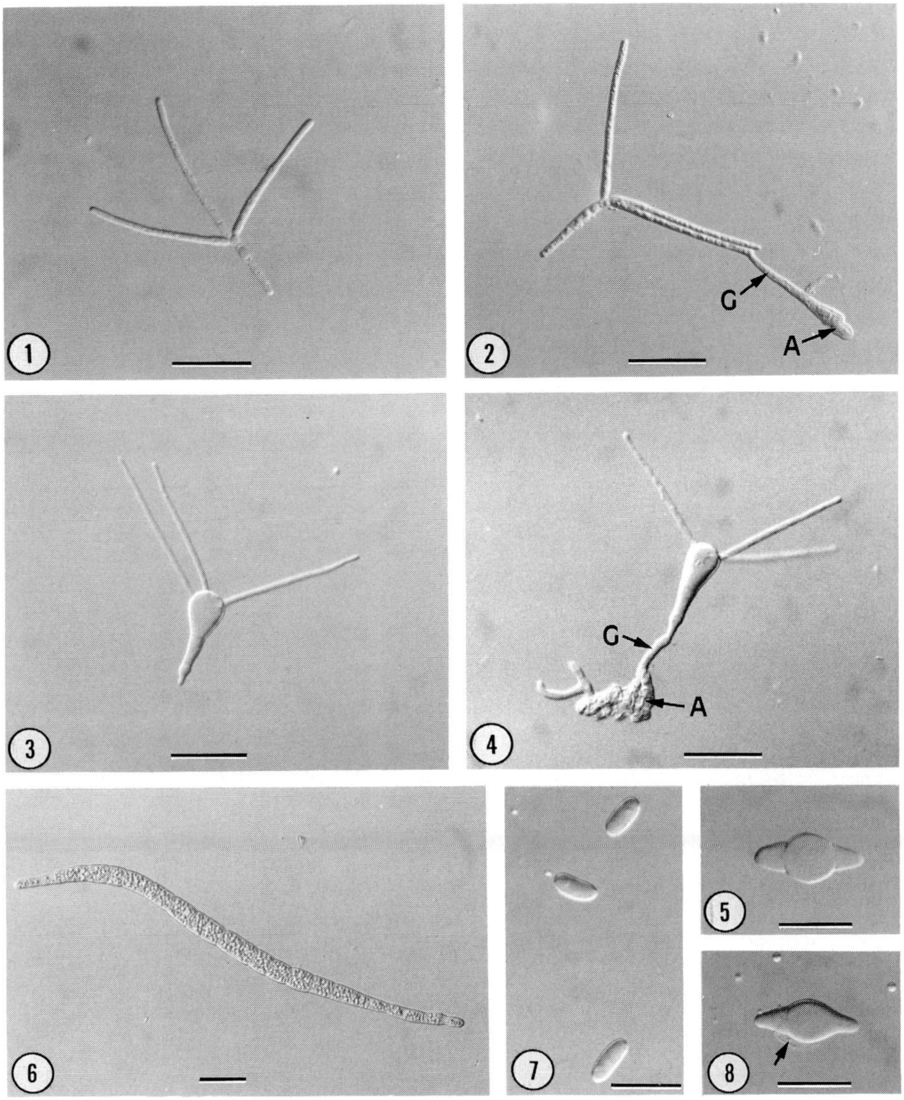

Fig. 7.1. 1–8 Light micrographs of settled conidia. **1** *Articulospora tetracladia*. **2** Conidium of *Articulospora tetracladia* with a short germ tube (*G*) and appressorium (*A*). **3** *Clavariopsis aquatica*. **4** Germinated conidium of *Clavariopsis aquatica* with a branched appressorium. **5** *Tumularia aquatica*. **6** *Anguillospora crassa*. **7** *Dimorphospora foliicola*. **8** Conidium of *Tumularia aquatica* with a sheet of mucilage (*arrowed*). *Bar* = 25 μm

and McLeod 1988). The percentage germination and germination rates of many of the aquatic hyphomycetes studied were either similar to, or higher than, those of the freshwater and marine aquatic fungi examined by Byrne and Jones (1975; Table 7.1).

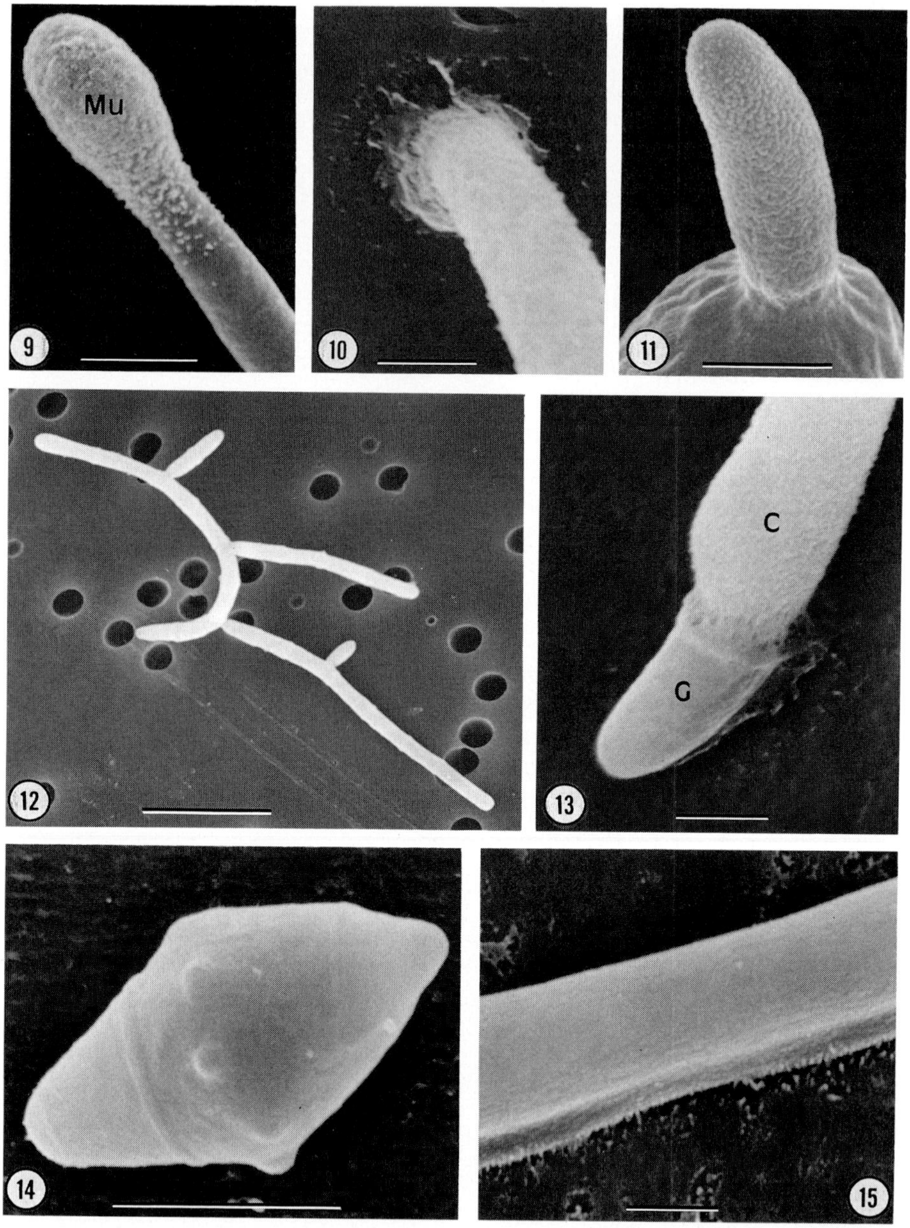

Fig. 7.2. 9–15 Scanning electron micrographs of settled conidia. **9** Swollen apex of the conidial arm of *Clavariopsis aquatica* with apical mucilage (*Mu*). **10** Mucilaginous pad at the apex of the conidial arm of *Articulospora tetracladia*. **11** Germ tube of *Dimorphospora foliicola*. **12** *Varicosporium elodeae*. **13** Conidium (*C*) of *Varicosporium elodeae* with a mucilage covered germ tube (*G*). **14** *Tumularia aquatica*. **15** Hyphal sheath of *Tumularia aquatica*. **9–11**, **13**, **15** *bar* = 2 μm; **12** *bar* = 20 μm; **14** *bar* = 10 μm

Table 7.1. Percentage germination of terrestrial, freshwater and marine fungi at 15 °C

Species	Germination (%)	Incubation (h)
Terrestrial[a]		
Mucor hiemalis Wehimer	16	24
Chaetomium globosum Kunze Ex Fr.	5	24
Penicillium notatum Westling	68	18
Freshwater[b]		
Anguillospora crassa Ingold	94	24
Articulospora tetracladia Ingold	98	24
Clavariopsis aquatica de Wild.	100	24
Dimorphospora foliicola Tubaki	100	6
Heliscus lugdunensis Sacc. et Therry	97	24
Lemonniera aquatica de Wild.	100	6
Mycocentrospora filiformis (Petersen) Iqbal	95	24
Tetracladium marchalianum de Wild.	97	24
Tetracladium setigerum[a] (Grove) Ingold	100	24
Tumularia aquatica (Ranzoni) Marvanová et Descals	88	24
Varicosporium elodeae Kegel	98	12
Marine[a]		
Asteromyces cruciatus (Moreau et Moreau) Ex Hennebert	72	42
Corollospora maritima Werdermann	100	24
Dendryphiella salina (Suth.) Pugh et Nicot	92	24
Zalerion maritimum (Linder) Anastasiou	78	24

[a] Byrne and Jones (1975).
[b] Read (1990).

It is concluded that the high germination rates of aquatic hyphomycete conidia contribute to their high competitive saprophytic ability compared to residual terrestrial fungi in the water and other aquatic competitors, particularly at the low temperatures characteristic of water courses in temperate zones. Rapid germination of spores at low temperatures by both freshwater and marine fungi indicates that this is a physiological adaptation common to many aquatic fungi.

The majority of the aquatic hyphomycetes germinated within 2–6 h. However, *Anguillospora crassa*, *Heliscus lugdunensis* and *Tumularia aquatica* germinated more slowly. These three species had other common features: their conidia contain large reserves of glycogen; show poor attachment to substrata; possess known teleomorphs, *Mollisia uda* Webster, *Nectria lugdunensis* Webster and *Massarina aquatica* Webster respectively (Webster and Descals 1979), which utilize woody substrata. A study of the colonization patterns of aquatic fungi on wood showed that following utilization of readily available sugars *H. lugdunensis* reverts to the *Nectria* teleomorph stage (Willoughby and Archer 1973). The large reserves of glycogen may enable conidia of these species to remain viable for long periods prior to utilization of extracellular nutrients. However, glycogen is metabolized slowly and this results in low germination rates and consequently slow initial attachment. Both rapid germination and attachment are required for colonization of leaf litter and so it is not surprising that these three species are found predominantly on wood. It is suggested that the morphologies of these conidia enable them to become entangled with, or lodged within, woody substrata and that rapid attachment is therefore less important.

The teleomorphs of many aquatic hyphomycetes remain unknown (Chap. 5). In *A. crassa*, *H. lugdunensis* and *T. aquatica*, it would appear that the teleomorph is the dominant stage and, as in many terrestrial fungi, the conidia are produced in large numbers only when conditions are favourable and serve to disperse the fungus rapidly.

7.4 Mucilage Secretion

Mucilage is secreted at the poles of the ungerminated arms of tetraradiate conidia either prior to or immediately subsequent to settlement (see Fig. 7.2, *9*, *10*; Fig. 7.4, *20*). In *Clavariopsis aquatica* a small amount of fibrillar mucilage is present on the apices of conidial arms prior to surface contact but is absent subapically (see Fig. 7.2, *9*), whereas in *Lemonniera aquatica* conidial mucilage is only secreted subsequent to settlement (Read 1990).

Movement of sigmoid conidia over a substratum is end over end with conidia arrested at one end (Webster and Davey 1984). The conidium finally attaches to the substratum at one pole and at one other region along its length. The production of mucilage for the purposes of attachment by ascospores of the marine fungus *Lulworthia medusa* (Ell. et Ev.) Cribb et Cribb shows a similar distribution pattern. In this case the ascospores are attached to substrata by drops of mucilage released from special end chambers (Rees and Jones 1984). Scanning electron microscope examination of sigmoid conidia of *Anguillospora crassa* and *Mycocentrospora filiformis* confirmed the presence of mucilage, produced initially at the points of contact with the substratum and then over the entire surface of the conidium (Read 1990). In turbulent water conidia may hit submerged objects many times and these multiple contacts may stimulate mucilage production. Ovoid conidia tend to produce a small amount of mucilage over their entire surface (see Fig. 7.4, *22*) prior to contact. *Tumularia aquatica* also produces a mucilaginous sheet beneath the conidium following settlement (see Fig. 7.1, *8*).

Mucilages produced by the aquatic hyphomycetes usually comprise polysaccharide, however, the nature and number of functional groups varies considerably between species (Read 1990). The mucilages of *A. tetracladia*, *C. aquatica* and *A. crassa* contain carboxyl, sulphate or phosphate functional groups. The mucilages of *T. aquatica* and *L. aquatica* comprise carboxyl and/or neutral groups, although *T. aquatica* also possesses a small proportion of sulphate and/or phosphate groups. The mucin/mucilage stains showed a negative reaction with *M. filiformis* which may indicate that the adhesive is a protein or glycoprotein.

At the SEM and TEM levels differences were noted between the texture of conidial and germ tube mucilages (Fig. 7.3, *17*, *18*; Fig. 7.4, *21–23*) see Fig. 7.2, *11*, *13*. This is particularly apparent for *Clavariopsis aquatica* in which three discrete extracellular mucilages are present. On the apices of conidial arms are two layers of fibrillar mucilage, subapically the mucilage is sparse or absent (see Fig. 7.2, *9*; Fig. 7.4, *20*, *21*) whereas the germ tubes are ensheathed by a compact electron-opaque mucilage (Fig. 7.4, *21*). The textural differences are paralleled by other differences. Conidial mucilages remained closely adpressed to the cell wall, whereas the germ tube mucilages and hyphal sheaths flowed extensively over the substratum to form hyphal sheaths (Read 1990) (see Fig. 7.2, *13*, *15*; Fig. 7.3, *17–19*; Fig. 7.4, *21*, *23–25*).

On recently fallen leaves which are not static, instantaneous attachment may not be critical. However, fallen leaves rapidly become submerged and trapped. In order for attachment to occur under these circumstances, the conidia must either become entrapped within the substratum, be adhesive, or produce an adhesive immediately upon contact. The possession of multiple contact points and rapid secretion of mucilage by tetraradiate conidia indicates that they are well adapted to the colonization of submerged leaf litter. The production of an adhesive at sites of contact with the substratum is a useful adaptation to the environment although it is not a mechanism unique to this ecological group of fungi.

The mechanism of mucilage release is not known, however a mechanism for triggering mucilage release must operate, as there has to be a critical time for its release. If the process is activated too early, then the effectiveness of the mucilage may be lost. Additional mucilage secretion is then initiated upon substratum contact which may result from either a physical or chemical response. Physical contact with a substratum causes a flattening of the cell wall. In algal carpospores this has been shown to create a calcium gradient within the spore and hyperpolarization of the basal spore membrane (Jones and Jones 1986). This was associated with a fundamental reassortment of the cell components with starch granules and vesicles containing adhesive material which migrated to the site of rhizoid formation. It is also possible that specific chemicals such as lectins may mediate in mucilage secretion. Lectins are produced by fungi from diverse taxa, however, many of these fungi are components of symbiotic associations and the lectins are involved in specific host recognition mechanisms (Nordbring-Hertz and Chet 1986). Owing to the non-specificity of conidial attachment in the aquatic hyphomycetes it is speculated that a physical factor triggers mucilage secretion and germination. Additionally, in the freshwater hyphomycetes *Anguillospora crassa* and *Lemonniera aquatica*, the nematode-trapping fungus *Arthrobotrys oligospora* Fresen. (Nordbring-Hertz 1988) and the marine ascomycete *Halosphaeria appendiculata* Linder (Hyde 1985),

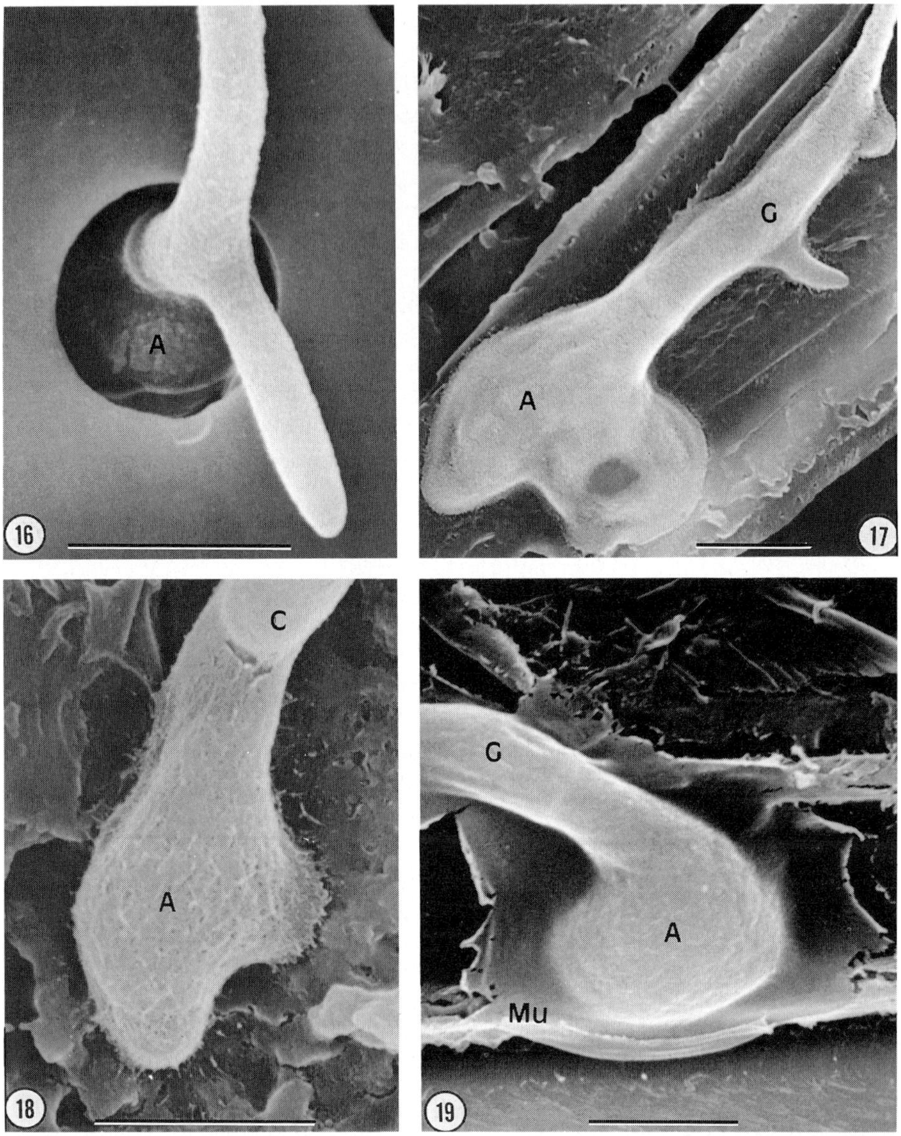

Fig. 7.3. 16–19 Scanning electron micrographs of appressoria. **16** Appressorium of *Mycocentrospora filiformis* within the pore of a nucleopore membrane. **17** Appressoria produced by *Mycocentrospora filiformis*; **18** *Lemonniera aquatica*; **19** *Varicosporium elodeae* settled on birch veneers. *A* Appressorium; *C* conidium; *G* germ tube; *Mu* mucilage; *Bar* = 5μm

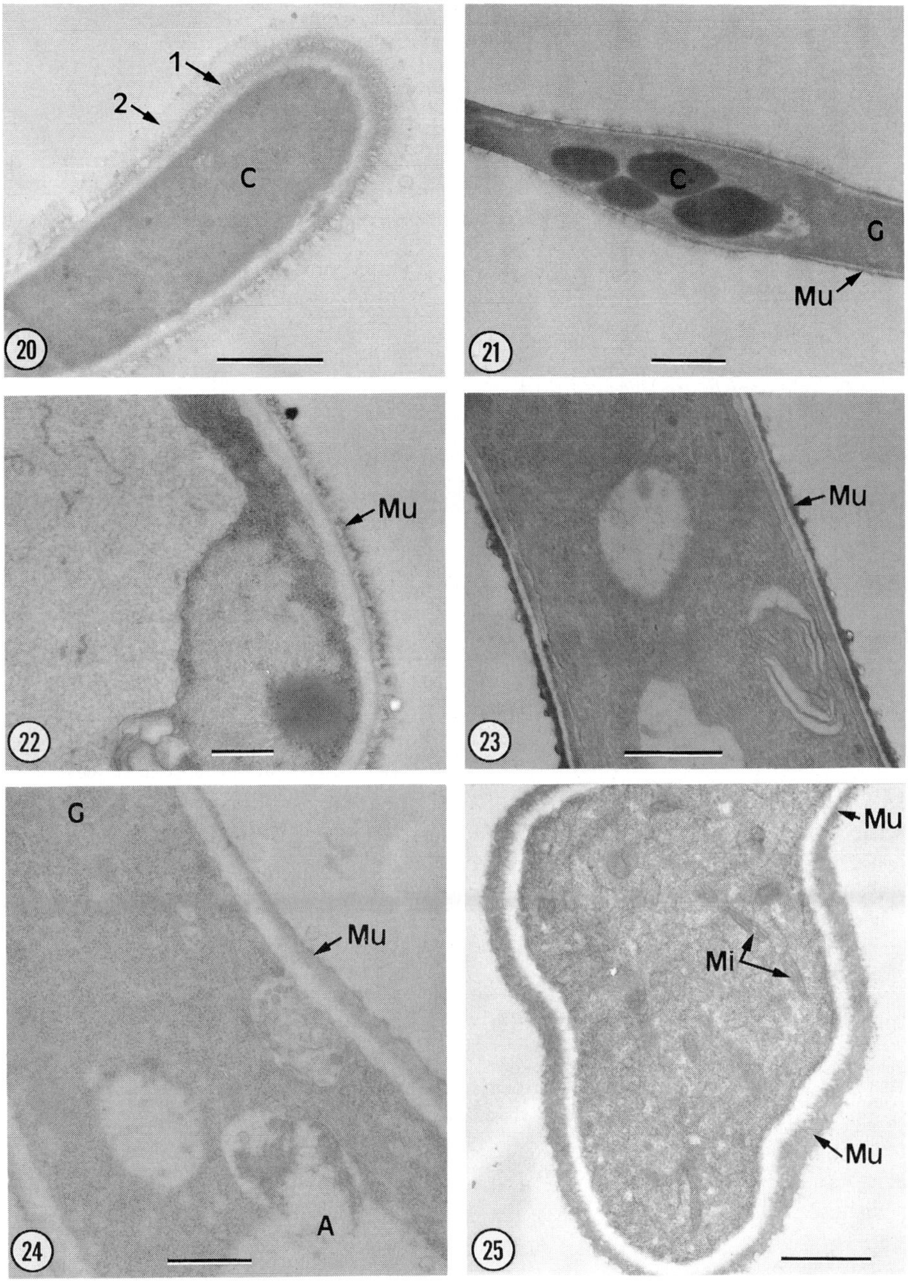

Fig. 7.4. 20–25 Transmission electron micrographs of settled conidia. **20**, **21**, **24** *Clavariopsis aquatica*. **20** Germinated conidium with a two-layered fibrillar mucilage at the apex of the conidial arm (*C*), an inner electron-opaque layer (*1*) and an outer electron lucent layer (*2*). A more compact electron-opaque mucilage (*Mu*) is present on the germ tube (*G*) (**21**) and appressorium (*A*) (**24**). **22**, **23** *Dimorphospora foliicola*. **22** Non-settled conidium with a layer of extracellular fibrillar mucilage (*Mu*). Note the increase in thickness of the extracellular mucilage (*Mu*) at the origin of the appressorium (*A*) of *Clavariopsis aquatica* (**24**) and *Lemonniera aquatica* (**25**). Note also the large number of mitochondria (*Mi*) in **25**. **20**, **21**, **25** *bar* = 1 μm; **22–24** bar = 500 nm

contact (or repeated contact) with the substratum is required before the fungus releases mucilage. This indicates the operation of a thigmotrophic mechanism for mucilage release.

7.5 Appressorium Formation

Mucilage-ensheathed appressoria were formed by all of the aquatic hyphomycete species examined with tetraradiate and branched conidia (see Fig. 7.1, *2*, *4*; Fig. 7.3, *18*, *19*; Fig. 7.4, *24*, *25*) and also by the sigmoid conidia of *Mycocentrospora filiformis* (see Fig. 7.3, *16*, *17*) (Harrison et al. 1988). Mucilage secretion has previously been reported for appressoria of other groups of fungi where it aids the attachment of the fungus to the substratum (Griffiths and Campbell 1973; Emmett and Parbery 1975; Corlett and Chong 1977). Ultrastructural evidence demonstrated that in the aquatic hyphomycetes, as in the plant pathogen *Colletotrichum atramentarium* (B. et Br.) Taub. (Griffiths and Campbell 1973), more mucilage is present on the appressoria than on the germ tubes and hyphae (see Fig. 7.3, *18*, *19*; Fig. 7.4, *24*, *25*). In *L. aquatica*, at the origin of the appressorium, there is a gradual increase in the thickness of the mucilage layer from 80 nm on the germ tube to 290 nm on the appressorium (see Fig. 7.4, *25*).

Appressoria are mostly observed in plant pathogens where their role is to produce an infection peg which then penetrates the host plant cell wall. In this instance one function of the appressorium is adhesion, sufficient to withstand the "back-pressure" created during penetration. In the saprophytic aquatic hyphomycetes, appressorial adhesion is required to prevent dislodgement of attached conidia and thus aid colonization.

Appressorium formation in the aquatic hyphomycetes was significantly affected by the substratum on which conidia settled. Firstly, appressoria were both more numerous and more complex on glass substrata than on nucleopore membranes (Read 1990). This concurred with the observations of Purkayastha and Menon (1981) on spores of *Colletotrichum corchori* Ikata et Tanaka in which appressoria were more numerous on a glass substratum than plant material. Appressorium complexity in plant pathogens (Tariq and Jeffries 1984) and the degree of spread and branching of algal rhizoids (Fletcher et al. 1984), are both directly related to the physical resistance of the substratum and its surface energy. In the case of aquatic hyphomycete species the differences in appressorium formation between substrata may also be accounted for by surface energy and hardness of the surfaces tested (Read 1990). The two are closely linked since hard surfaces frequently have high surface energy. This may be due to the surface energy or the presence of specific topographical (Preece and Dickinson 1971; Whiteside 1973) or chemical (Waller 1970; Whiteside 1974; Emmett and Parbery 1975) stimuli. Appressoria of the aquatic hyphomycetes may have one or more roles:

1. To enable greater attachment to substrata by an increased area of contact and secretion of mucilage. In species with tetraradiate and branched conidia the appressoria are frequently produced at the tips of the arms immediately upon germination (e.g. *Tetrachaetum elegans* Ingold, Webster 1959; *Lemonniera aquatica*, Read 1990). The rapid adhesion of tetraradiate and branched conidia argues for a primary role in attachment for appressoria in these forms. The formation of appressoria on extended germ tubes and hyphae indicates that they may also adhere vegetative hyphae to substrata.
2. Willoughby and Hasenjäger (1987) suggested that in *Saprolegnia* spp. appressoria may act as assimilatory organs. As aquatic hyphomycetes produce a wide range of extracellular hydrolytic enzymes (Zemek et al. 1985) and their appressoria possess both high levels of metabolic activity (see Fig. 7.4, *25*) and mucilaginous sheaths (Read 1990) it is possible that appressoria in aquatic hyphomycete species may have an assimilatory function.
3. To enable the production of an infection structure. Penetration of the substratum has not been observed (Read 1990). However, it cannot be discounted, particularly in *Mycocentrospora filiformis*, as a related, but not obligately aquatic, species, *M. acerina*, does produce pentration pegs from appressoria (Wall and Lewis 1980).

7.6 Influence of Attachment Structures on the Strength of Adhesion

A similar sequence of attachment has been identified in the aquatic hyphomycetes as was indicated by Jones et al. (1983) for carpospores of the macroalga, *Ceramium rubrum* L. A comparison of the attachment stages of aquatic hyphomycetes and *Ceramium rubrum* is presented in Table 7.2. A more detailed account of the attachment of aquatic hyphomycetes with regard to their conidium configurations is detailed below.

Table 7.2. Comparison of the ontogeny of mucilage production and germination structures by carpospores of *Ceramium rubrum* and aquatic hyphomycete conidia

Stages of development	*Ceramium rubrum*	Aquatic hyphomycetes
Initial settlement	Peripheral mucilage	Peripheral mucilage
2 h (spore attachment)	Specific released carpospore adhesive	Conidial mucilage
6–12 h (germination)	Primary rhizoid formation	Germ tube formation
12–24 h (attachment structures)	Filamentous and discoid rhizoids	Appressoria[a]
24 h + (maximum attachment)	Extensive rhizoid networks	Complex appressoria[a]/ vegetative mycelium

[a] Formed by all tetraradiate and branched conidia and one of the sigmoid conidia.

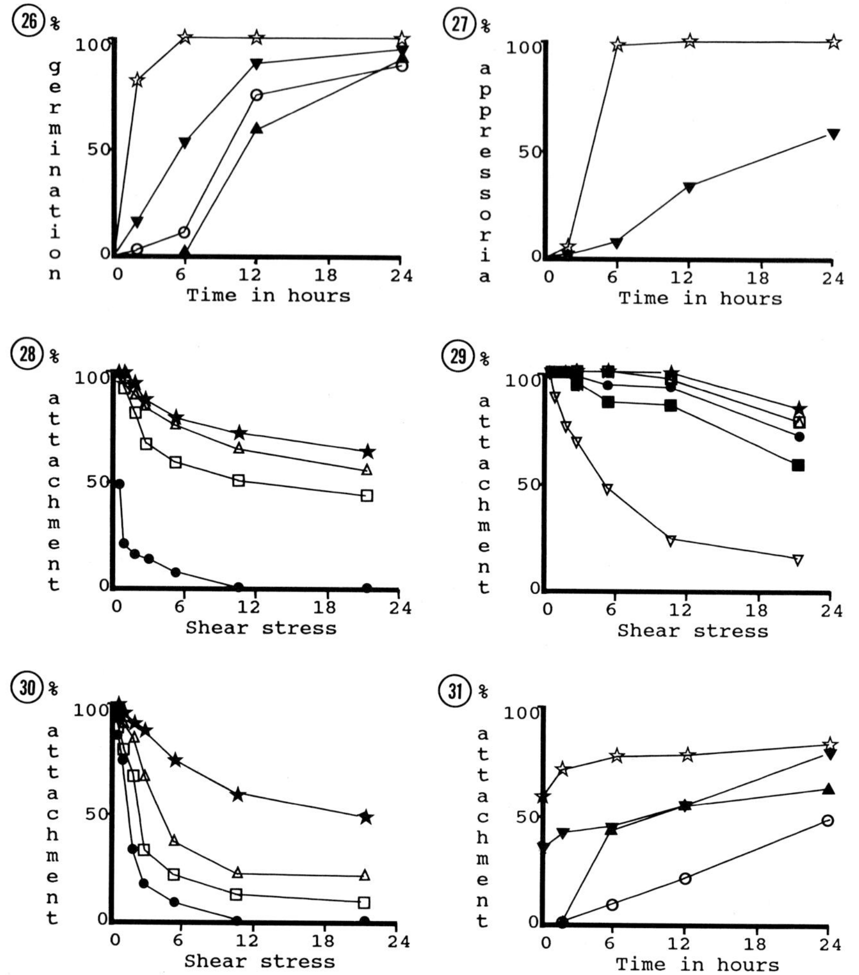

Fig. 7.5. **26**, **27** Development of germ tubes and appressoria. **26** Percentage germination of conidia of aquatic hyphomycetes up to 24 h after settlement. **27** Percentage of germinated conidia which had formed appressoria. ☆ = *Lemonniera aquatica*, ▼ = *Mycocentrospora filiformis*, ○ = *Tumularia aquatica*, ▲ = *Anguillospora crassa*. **28–30** Attachment of conidia of aquatic hyphomycetes to perspex discs after settlement for up to 24 h and subjection to a shear stress of up to 21.26 Nm^{-2} for 5 min. **28** *Anguillospora crassa*; **29** *Lemonniera aquatica*; **30** *Tumularia aquatica*. ▽ = 5 min, ■ = 30 min, ● = 2 h, □ = 6 h, △ = 12 h, ★ = 24 h. **31** The effect of mucilage secretion, germination and appressorium formation on the strength of attachment of conidia of aquatic hyphomycetes to perspex discs after settlement for 24 h and subjection to a shear stress of 21.26 Nm^{-2} for 5 min. ☆ = *Lemonniera aquatica*, ▼ = *Mycocentrospora filiformis*, ○ = *Tumularia aquatica*, ▲ = *Anguillospora crassa*

7.6.1 Tetraradiate Conidia

In tetraradiate and branched conidia the general patterns of germination and appressorium development were similar and the initial conidium/substratum contact was maintained by mucilage. After settlement for 6–12 h further secretion of mucilage by the conidia and the presence of germ tubes with mucilaginous sheaths resulted in increased attachment. Appressoria developed approximately 12 h after settlement. At high shear stresses, marked increases in attachment occurred 24 h after settlement which correlated with the presence of mature appressoria.

Conidia of *L. aquatica* showed the most rapid development. Mucilage was secreted within 5 min of settlement, germination occurred within 30 min (Fig. 7.5, *26*) and protoappressoria formed after 2 h (Fig. 7.5, *27*). Attachment studies showed that conidia were strongly attached within 2 h (Fig. 7.5, *29*) and that the strength of adhesion was correlated with the development of the different attachment structures (Fig. 7.5, *26*, *27*). The formation of complex, multiple appressoria and the "germination" of the appressoria giving rise to hyphae after 24 h resulted in further increases in adhesion. It is therefore concluded that the extremely strong attachment of *L. aquatica* is due to its rapid germination rate and the subsequent rapid production of adhesive pads and appressoria.

7.6.2 Sigmoid Conidia

Conidia of *Mycocentrospora filiformis* are more securely attached to a substratum after settlement for 5 min than those of *A. crassa* after 2 h (Fig. 7.5, *31*). Germination had not occurred at this stage (Fig. 7.5, *26*) and the two factors which influenced the strength of attachment were: (1) the amount and efficacy of the mucilage present on the conidia; (2) conidial dimensions. It was shown that the mucilage composition of the two species was different. The mucilage of *A. crassa* is a polysaccharide whereas that of *M. filiformis* is thought to comprise protein or glycoprotein. The latter compound may be a more powerful adhesive. The second factor is rather complex. In polar view sigmoid conidia present similar profiles, however, conidia of *A. crassa* possess a greater cross-sectional area than those of *M. filiformis* and therefore create higher resistance to water currents which may result in their greater dislodgement.

Between 2 and 6 h after settlement the secretion of conidial mucilage resulted in attachment of conidia of *A. crassa* to the substratum (Fig. 7.5, *28*). Further secretion of mucilage and the formation of germ tubes (Fig. 7.5, *26*) resulted in increased attachment between 6 and 12 h after initial settlement (Fig. 7.5, *28*). The influence of appressorium formation on the strength of attachment is demonstrated by these two species. Formation of appressoria by *M. filiformis* 12 h after settlement significantly increases its strength of attachment whereas *A. crassa* without appressoria showed only a slight increase (Fig. 7.5, *31*).

It is concluded that the conidia of *A. crassa* do not attach well to smooth surfaces at high flow rates. However, they might attach more readily to uneven surfaces by becoming entrapped within the substratum. In the natural environment, the flow rates and hence the shear stresses imposed on newly settled conidia within the laminar boundary layer might be much lower than the experimental conditions, enabling the conidia to attach more efficiently.

7.6.3 Ovoid Conidia

None of the ovoid conidia attached rapidly (Fig. 7.5, *30*). This may be due to the high ratio of conidium volume to the area of contact with the substratum. As development proceeded with the production of germ tubes, hyphae and hyphal sheaths, the increases in the strength of attachment were proportional to time and once established the conidia remained securely attached to the substratum even at high shear stresses. In *Dimorphospora foliicola* this was as strong as in the tetraradiate conidia with their more highly adapted conidium morphology (Read et al. 1988).

7.7 Conclusions

This chapter has reviewed our knowledge of the settlement and attachment of aquatic hyphomycetes to surfaces. The following conclusions are drawn.

1. The prime factor which influences attachment is conidium morphology. Tetraradiate and branched conidia show the best initial attachment for they have at least three points of contact with the substratum, while ovoid and sigmoid conidia only possess one or two contact points.
2. Conidia produce mucilage at the region(s) most likely to make initial contact with a substratum, e.g. tips of the conidia. The mucilage may be produced as the result of a contact response to the substratum or may have been formed and released prior to contact.
3. Rapid germination rates combined with mucilage production by the germ tubes account for the increases in adhesion observed between 2 and 12 h after settlement and attachment.
4. The formation of appressoria by tetraradiate and branched conidia and by the sigmoid conidia of *Mycocentrospora filiformis* is correlated with a significant increase in the strength of attachment.
5. Mucilaginous hyphal sheaths are ubiquitous and aid in attachment prior to colonization of the substratum.

It is proposed that rapid attachment of aquatic hyphomycete conidia is a major adaptation to the aquatic environment although further work on the origin and site of mucilage production, its transportation to the plasma-membrane and the mechanism(s) which trigger mucilage release is required.

Acknowledgements. We would like to thank Prof. J. Webster for the culture of *Dimorphospora foliicola*, Mr. C. Derrick for photographic assistance and the N.A.B. for financial support.

References

Byrne P, Jones EBG (1975) Effect of salinity on spore germination of terrestrial and marine fungi. Trans Br Mycol Soc 64:497–503

Chamier A-C, Dixon PA, Archer SA (1984) The spatial distribution of fungi on decomposing alder leaves in a freshwater stream. Oecologia 64:92–103

Corlett M, Chong J (1977) Ultrastructure of the appressorium of *Spilocaea pomi*. Can J Bot 55:5–7

Cox PA (1983) Search theory, random motion, and the convergent evolution of pollen and spore morphology in aquatic plants. Am Nat 121:9–31

Emmett RW, Parbery DG (1975) Appressoria. Annu Rev Phytopathol 13:147–167

Fletcher RL, Baier RE, Fornalik MS (1984) The influence of surface energy on spore development in some common marine fouling algae. Proc 6th Int Congr on Marine corrosion and fouling, marine biology, Comité-International Permanent pour la Recherche sur la Préservation des Matériaux en Milieu Marin, Athens, Greece, pp 129–144

Fowler HW, McKay AJ (1980) The measurement of microbial adhesion. In: Berkeley RCW, Lynch JM, Melling J, Butler PR, Vincent B (eds) Microbial adhesion to surfaces. Ellison Horwood, Chichester, pp 143–161

Griffiths DA, Campbell WP (1973) Fine structure of conidial germination and appressorial development in *Colletotrichum atramentarium*. Trans Br Mycol Soc 61: 529–536

Harrison SJ, Moss ST, Jones EBG (1988) Fungal adhesion in aquatic hyphomycetes. Int Biodeterior Bull 24:271–276

Hyde KD (1985) Spore settlement and attachment in marine fungi. Thesis, Portsmouth Polytechnic, UK

Hyde KD, Moss ST, Jones EBG (1989) Attachment studies in marine fungi. Biofouling 1:287–298

Ingold CT (1953) Dispersal in fungi. Oxford University Press, Oxford

Ingold CT (1954) Presidential address. Fungi and water. Trans Br Mycol Soc 37:97–107

Ingold CT (1956) Stream spora in Nigeria. Trans Br Mycol Soc 39:108–110

Ingold CT (1966) The tetraradiate aquatic fungal spore. Mycologia 58:43–56

Ingold CT (1975) An illustrated guide to aquatic and water-borne Hyphomycetes (Fungi Imperfecti) with notes on their biology. Freshwater Biol Assoc Sci Pub No 30, 96 pp

Ingold CT (1976) The morphology and biology of freshwater fungi excluding Phycomycetes. In: Jones EBG (ed) Recent advances in aquatic mycology. Elek Science, London, pp 335–357

Ingold CT (1984) Aquatic Hyphomycetes with tetraradiate conidia. In: Taxonomy of fungi. Proc Int Symp on Taxonomy of fungi. University of Madras, vol 2:353–364

Iqbal SH, Webster J (1973) The trapping of aquatic hyphomycete spores by air bubbles. Trans Br Mycol Soc 60:37–48

Jones AM, Jones EBG (1986) Studies of the marine fouling alga *Ceramium rubrum*. I. Polarization of settled carpospores. In: Barry S, Houghton DR (eds) Proc 6th Int Biodeterioration Symp. CAB International, Slough, pp 590–595

Jones AM, Fletcher RL, Daniel GF, Jones EBG (1983) Settlement and adhesion of algae. In: Mauchline J (ed) Fouling and corrosion of metals in sea water. Scottish Marine Biological Association, Oban, pp 31–77

Magan N, McLeod AR (1988) In vitro growth and germination of phylloplane fungi in atmospheric sulphur dioxide. Trans Br Mycol Soc 90:571–575

Nordbring-Hertz B (1988) Ecology and recognition in the nematode-nematophagous fungus system. In: Marshall KC (ed) Advances in microbial ecology, vol 10. Plenum, New York, pp 81–114

Nordbring-Hertz B, Chet I (1986) Fungal lectins and agglutinins. In: Mirelman D (ed) Microbial lectins and agglutinins: properties and biological activity. Wiley and Sons, New York, pp 393–408

Preece TF, Dickinson CH (1971) Ecology of leaf surface micro-organisms. Academic Press, London

Purkayastha, RP, Menon U (1981) Factors affecting appressorium formation by *Colletotrichum corchori*. Trans Br Mycol Soc 77:183–185

Read SJ (1990) Spore attachment in fungi with special reference to freshwater Hyphomycetes. Thesis, Portsmouth Polytechnic, UK

Read SJ, Moss ST, Jones EBG (1988) Establishment of the hyphomycete biofilm. In: Morton LHG, Chamberlain AHL (eds) Biodeterioration Society, Occasional Publication No 4, Biofilms. Biodeterioration Society Kew, UK, pp 88–96

Rees G, Jones EBG (1984) Observations on the attachment of spores of marine fungi. Bot Mar 27:145–160

Sanders PF, Webster J (1980) Sporulation responses of some "aquatic Hyphomycetes" in flowing water. Trans Br Mycol Soc 74:601–605

Shearer CA, Webster J (1985a) Aquatic hyphomycete communities in the River Teign. I. Longitudinal distribution patterns. Trans Br Mycol Soc 84:489–501

Shearer CA, Webster J (1985b) Aquatic hyphomycete communities in the River Teign. II. Temporal distribution patterns. Trans Br Mycol Soc 84:503–507

Suberkropp K (1984) Effect of temperature on seasonal occurrence of aquatic hyphomycetes. Trans Br Mycol Soc 82:53–62

Tariq VN, Jeffries P (1984) Appressorium formation by *Sclerotinia sclerotiorum*: scanning electron microscopy. Trans Br Mycol Soc 82:645–651

Wall CJ, Lewis BG (1980) Infection of carrot leaves by *Mycocentrospora acerina*. Trans Br Mycol Soc 75:163–165

Waller JM (1970) Sugarcane smut (*Ustilago scitaminea*) in Kenya. II. Infection and resistance. Trans Br Mycol Soc 54:405–414

Webster J (1959) Experiments with spores of aquatic hyphomycetes. I. Sedimentation and impaction on smooth surfaces. Ann Bot 23:595–611

Webster J (1981) Biology and ecology of aquatic hyphomycetes. In: Wicklow DT, Carroll GC (eds) The fungal community, its organisation and role in the ecosystem. Marcel Dekker, New York, pp 681–691

Webster J (1987) Convergent evolution and the functional significance of spore shape in aquatic and semi-aquatic fungi. In: Rayner ADM, Brasier CM, Moore D (eds) Evolutionary biology of the fungi. Cambridge University Press, Cambridge, pp 191–201

Webster J, Davey RA (1984) Sigmoid conidial shape in aquatic fungi. Trans Br Mycol Soc 83:43–52

Webster J, Descals E (1979) The teleomorphs of water-borne hyphomycetes from freshwater. In: Kendrick B (ed) The whole fungus, vol 2. National Museum of Natural Sciences, Ottawa, pp 419–451

Webster J, Descals E (1981) Morphology, distribution and ecology of conidial fungi in freshwater habitats. In: Cole GT, Kendrick B (eds) Biology of conidial fungi, vol 1. Academic Press, London, pp 295–352

Whiteside JO (1973) Action of oil in the control of citrus greasy spot. Phytopathology 63:262–266

Whiteside JO (1974) Environmental factors affecting infection of citrus leaves by *Mycosphaerella citri*. Phytopathology 64:115–120

Willoughby LG, Archer JF (1973) The fungal spora of a freshwater stream and its colonization pattern on wood. Freshwater Biol 3:219–239

Willoughby LG, Hasenjäger R (1987) Formation and function of appressoria in *Saprolegnia*. Trans Br Mycol Soc 89:373–380

Zemek J, Marvanová L, Kuniak L, Kadlecikova B (1985) Hydrolytic enzymes in aquatic hyphomycetes. Folia Microbiol 30:363–372

8 Water Chemistry

Anne-Carole Chamier

8.1 Introduction

Limnologists have long recognized the importance of water chemistry on the distribution and physiology of aquatic biota, especially of photosynthetic organisms and aquatic animals in autotrophic ecosystems (Hynes 1970). The effects of water chemistry on heterotrophic ecosystems are coming to light largely through investigations of acid deposition. Most of these have concentrated on the relationship between pH and salmonid fish populations, some on aquatic invertebrates, but few on the aquatic fungi and bacteria which degrade allochthonous litter – a major source of energy for all trophic levels of the ecosystem. Whilst pH is an important factor characterizing water chemistry, biologically important nutrients and inhibitors are not necessarily dependent on pH. The aim of this review is to demonstrate how little is known about water chemistry and aquatic hyphomycetes, the dominant degradative fungi in freshwater streams, and to suggest further lines of research.

8.2 Field Studies – Species Diversity and pH

In early field studies of aquatic hypomycete species in Japanese lakes, Suzuki and Nimura (1960a,b; 1961) found significantly higher numbers of species in harmonic (pH 6.2–6.5) than in acidotrophic waters (pH 3.8–5.8). Later, Iqbal and Webster (1973, 1977) found that rivers with higher alkalinity and conductivity (pH 6.0–9.0) carried twice the conidial concentration and numbers of species of low alkalinity, low conductivity, moorland streams (pH 4.1–6.45). The River Teign in Devon traverses both upland (pH 5.4–6) and lowland (pH 7–7.2). Shearer and Webster (1985) reported that the upstream waters had a distinct and, compared to the downstream sites, impoverished fungal flora. Chamier (1987) investigated leaf degradation in seven streams of the English Lake District (pH 4.9–6.8). On oak leaves in upland streams there were three to six species of aquatic hyphomycetes regardless of pH (4.9–6.8) or vegetation, compared to 14 species in two lowland streams of pH 6.8; one with trees, the other treeless. However, surveys of 26 wooded streams in Europe and Canada by Wood-Eggenschwiler and Bärlocher (1983) and Bärlocher (1987) found that species number remained fairly consistent between pH 5–7, declining slightly above and below these values. Further evidence that species diversity is highest in circumneutral streams is provided by Marvanová (1984) and Regelsberger et al. (1987).

These studies of species diversity in relation to pH are not strictly comparable as there is no consistency overall in the time period over which sampling was carried out, the frequency and number of samples or the source of samples. Table 8.1 sets out the relevant data on species diversity in streams of contrasting pH and water chemistry where one set of workers have used consistent techniques and sample sources.

8.3 Alkalinity and pH

A critical factor in the pH chemistry of freshwaters is the level of alkalinity. This is the sum of anions of weak acids, hydroxyl ions and bicarbonate. At a pH > 6 bicarbonate ions (HCO_3^-) are usually the dominant component. The pH rises with increasing bicarbonate, but is also influenced by the concentration of CO_2. At pH 5.7, alkalinity is around zero and waters normally below this value have little buffering capacity. A pH of 5.7 has been found to be a critical value for the survival of many aquatic invertebrates (Sutcliffe and Carrick 1973). It seems to be an important threshold for all biota and discussion of pH tolerances in freshwaters is aided if an average pH < 5.7 is termed acidic, pH > 5.7–7.2 circumneutral and pH > 7.2 alkaline.

8.3.1 Studies in Acidic Streams

By this terminology, which depends upon data from frequent, regular sampling, few studies of aquatic hyphomycete communities have been made in acidic waters, viz. the upland, Dartmoor streams investigated by Iqbal and Webster (1977); the upland Walla Brook on the River Teign (Shearer and Webster 1985); a Canadian woodland stream, Allen Creek, sampled by Bärlocher (1987); three upland streams in the Lake District (Chamier 1987). Characteristic of all these streams were very low levels of conductivity (<50 μmho), alkalinity and calcium ion concentration ($<8\,mg\,l^{-1}$). However, whilst all upland, moorland streams, regardless of pH, were found to have a very low number of aquatic hyphomycete species, the low pH wooded stream, Allen Creek, had species numbers comparable with circumneutral streams (Table 8.1). It might be concluded that site and substrate availability are as important determinants of aquatic hyphomycete communities as pH.

Table 8.1. Species diversity of aquatic hyphomycetes in acid, circumneutral and alkaline streams

Stream type	Name	Alkalinity (meq l^{-1})	Ca^{2+} (mg l^{-1})	pH	Number of species	Source of fungi
a) Iqbal and Webster (1977)						
Upland acidic	Cherry Brook	0.10	2	4.4–6.4	21	Stream water
	R. Cowsic	0.08–0.1	1.6–2.0	4.0–6.1	11	
	R. Swincome	0.08–0.1	1.6–2.0	4.2–6.4	12	
Iqbal and Webster (1973)						
Lowland circumneutral to alkaline	R. Culvery	0.64	25.6	6.9–8.0	27	Stream water
	R. Creedy	1.54	30.8	6.4–9.0	34	
	R. Exe	1.00	20	6.3–8.7	24	
b) Shearer and Webster (1985)						
Upland acidic	Walla Brook (R. Teign site I)	?	?	5.4–6.0	21	Stream water
Lowland circumneutral	R. Teign (site II)	?	?	7.0	37	Stream water
	R. Teign (site III)	?	?	7.2	37	
c) Chamier (1987)						
Upland acidic	Gaitscale Gill	−0.01	2.6	4.9	4	Oak leaf packs
	Mosedale Beck	−0.01	2.7	5.0	4	
	Wrynose Bottom	0.005	4.3	5.5	3	
Upland circumneutral	Hardknott Gill	0.067	7.9	6.6	3	
	Carole's Beck	0.110	8.5	6.8	6	
Lowland circumneutral	Upper Crosby	0.220	14.2	6.8	14	
	Lower Crosby	0.182	13.2	6.8	14	

d) Bärlocher (1987)						
Acidic	Allen Creek	0.10	1.5	5.4	40	Foam
Circumneutral	8 Canadian streams	0.10–0.90	1.8–16.0	5.9–7.2	x = 34	
Alkaline	Cedar Camp Brook	0.50	12.9	7.4	31	
e) Wood-Eggenschwiler and Bärlocher (1983)						
	Black Forest					
Circumneutral	Wehra	0.32	7.5	6.97	32	Water, foam, leaves
	Hofwald	0.17	5.5	6.09	41	
	Vosges					
Alkaline	Steinbyruntz	0.76	12.8	7.5	41	
	Siehl	0.49	8.5	7.3	29	
	Jura					
	Neuenstein	4.32	79.7	8.1	28	
	Fehren	4.50	81.7	8.0	28	
	Napf					
	Luthern	3.70	70.1	8.3	26	
	Chuttelbad	3.33	64.6	8.4	26	

Table 8.2. Weight loss coefficient k and days taken to 50% weight loss (t_{50}) of alder and oak leaves in circumneutral and acid streams (Chamier 1987)

Stream	pH	k		t_{50}	
		Alder	Oak	Alder	Oak
Upper Crosby	6.8	−0.0164	−0.00436	42	159
Lower Crosby	6.8	−0.0131	−0.00333	53	208
Carole's Beck	6.8	−0.0119	−0.00333	58	208
Hardknott Gill	6.6	−0.0101	−0.00319	69	217
Wrynose Bottom	5.5	−0.0031	−0.00246	223	282
Mosedale Beck	5.0	−0.0030	−0.00244	233	284
Gaitscale Gill	4.9	−0.0032	−0.00283	215	245

The acidic Dartmoor streams with no riparian trees have a distinctive assemblage of species, although a few are known to be ubiquitous, e.g. *Varicosporium elodeae*, *Articulospora tetracladia*, *Clavatospora longibrachiata*, *Alatospora acuminata*, *Tricladium giganteum*. These species were found in acidic streams in the Lake District and included *Alatospora constricta* which was found only at upland sites.

If low pH in itself is not the sole factor accounting for low species numbers and conidial concentration of aquatic hyphomycetes, the chemistry of low pH waters nevertheless appears to affect fungal metabolism, even in species that are acid-tolerant. Studies of the rates of degradation of leaves in streams show that breakdown is significantly slower in acid than in circumneutral streams (Friberg et al. 1980; Mackay and Kersey 1985; Chamier 1987; Mulholland et al. 1987). Only one of these studies followed the course of aquatic hyphomycetes colonizing decomposing leavels. Leached alder leaves (*Alnus glutinosa* (L.) Gaertn.) were degraded by microbes four times more rapidly in circumneutral than in acidic streams (Table 8.2). Total cumulative counts of aquatic hyphomycete conidia made from strips cut from submersed leaves showed a common value from circumneutral streams associated with 2–11 species. Total cumulative conidial counts from acidic streams were one tenth of the circumneutral values and associated with four to five species (Fig. 8.1). Initial levels of colonization on leaves in acidic streams were comparatively high but showed a slow, consistent decline later on.

Oak leaves (*Quercus robur* L.) were degraded twice as rapidly in circumneutral as in acidic streams (Table 8.2). Cumulative conidial counts followed the pattern for alder leaves, but cumulative species counts were about 16 in lowland circumneutral streams and about four in upland, compared to four to six species in acidic streams (Fig. 8.1). The indications are that species diversity is not as important in determining rates of decomposition as factors associated with low pH water chemistry that adversely affect fungal metabolism.

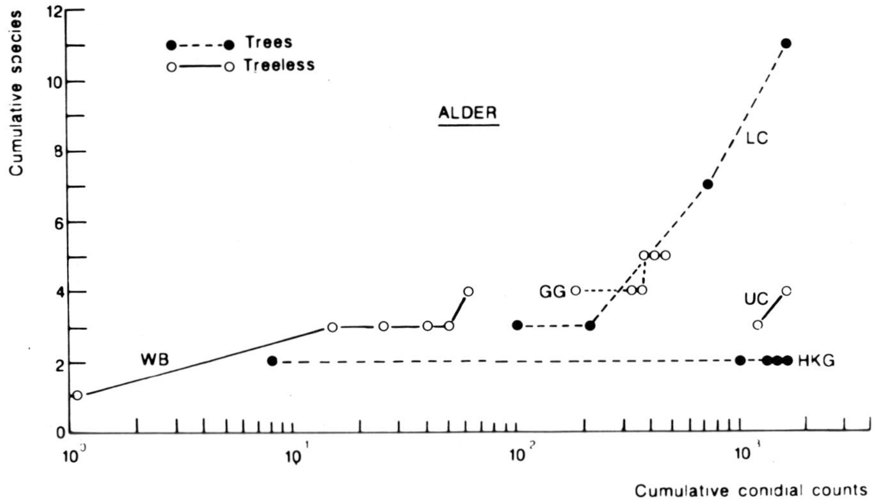

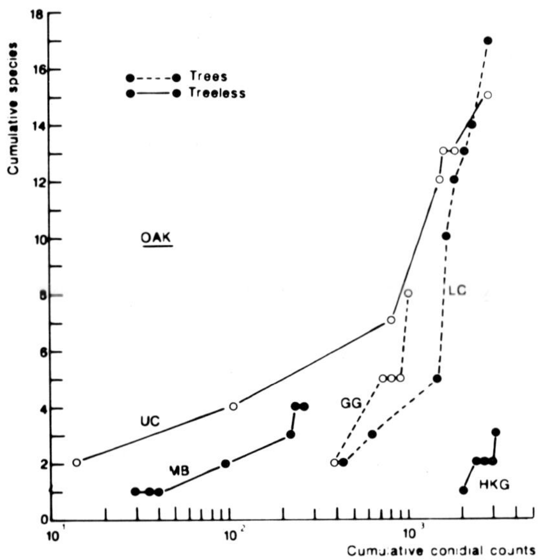

Fig. 8.1. Cumulative conidial counts versus cumulative aquatic hyphomycete species from alder (*top*) and oak leaf samples submersed in circumneutral (*LC*, *UC*, *HKG*) and acid (*GG*, *WB*, *MB*) streams. Data from Chamier (1987)

8.3.2 Chemical Characteristics of Acid Waters

The toxicity of acidic waters to salmonid fish and invertebrates has been attributed largely to the effects of high concentrations of soluble, monomeric and other free forms of inorganic aluminum (Al_i) combined with low concentrations of calcium ions. These conditions exist where stream catchment areas lie over acidic, base-poor soils and rocks. Very heavy rainfall in these situations with overland flow or interflow (Hynes 1983) can result in a considerable drop in pH with increases in inorganic aluminum (Al_i) levels through disturbance of stream sediments and run-off from acidified upper soil horizons. Similar effects occur at snow-melt. These are the most extreme, short-term acidic episodes. Normally, when rainwater seeps into soils, free aluminum is released at pH 5 by cationic exchange with hydrogen ions.

In base-rich soils, mobilization of alkaline cations in the presence of organic acids and bicarbonate creates a buffered waterbody that can neutralize strong acids and stabilize pH. Toxic forms of aluminum in streams and lakes fed from base-rich soils and bed-rocks appear to complex with organic acids, silicic acid and inorganic ligands to form non-toxic species. Complexed aluminum, high calcium-ion concentration, high buffering capacity and high conductivity are characteristics of circumneutral and alkaline streams.

Below a pH of 5.7, where seepage is from acidic, leached soils, free monomeric aluminum occurs at high levels; ionic concentrations, particularly of calcium, are very low; the water has very poor buffering capacity, and acidity is due to strong acids. It is this combination of features that is biologically harmful. Where streams have high concentrations of organic acidity and there is a low fraction of labile, monomeric aluminum, low pH in itself does not appear to be as damaging to the biota, even where calcium-ion concentration and conductivity are low (Kerekes 1989; Sutcliffe 1983).

These distinctions in low pH water chemistry may help to explain the significant differences in aquatic hyphomycete flora between the wooded acidic streams investigated by Bärlocher (1987) and Bärlocher and Petersen (1988) with rich populations and the impoverished upland, moorland streams studied by Iqbal and Webster (1977), Shearer and Webster (1985) and Chamier (1987) (see Table 8.1). Deleterious effects would be magnified for aquatic hyphomycetes by the fact that concentrations of most ions vary in a pronounced seasonal pattern with inorganic aluminum levels three times higher in winter than in summer, and calcium ion concentration lowest in winter (Tipping 1990).

8.3.3 Elevated Aluminum (Al_i) Levels in Acidic Streams and Reduced Microbial Metabolism on Litter

The implication of aluminum and calcium ion concentrations for reduced microbial metabolism on allochthonous litter in acidic streams is suggested by two studies. Mulholland et al. (1987) found that leaf decomposition rates were significantly lower in streams with a pH < 5.7 compared with a stream of a pH of 6.4. Differences between streams of pH 4.5–5.7 were not significant. Microbial ATP and respiration rates followed the same pattern as weight loss in leaf packs. These workers concluded that acidic waters lowered rates of microbial metabolism. Palumbo et al. (1987) analyzed aluminum levels of leaves submersed in an acidic (pH 4.9) and a circumneutral (pH 6.4) stream and found about three times more aluminum on the leaves in the acidic stream. Transfer of leaves from the high to the low pH stream reduced microbial respiration within a week. A reverse transfer from low to high pH produced significantly positive results after only 4 weeks.

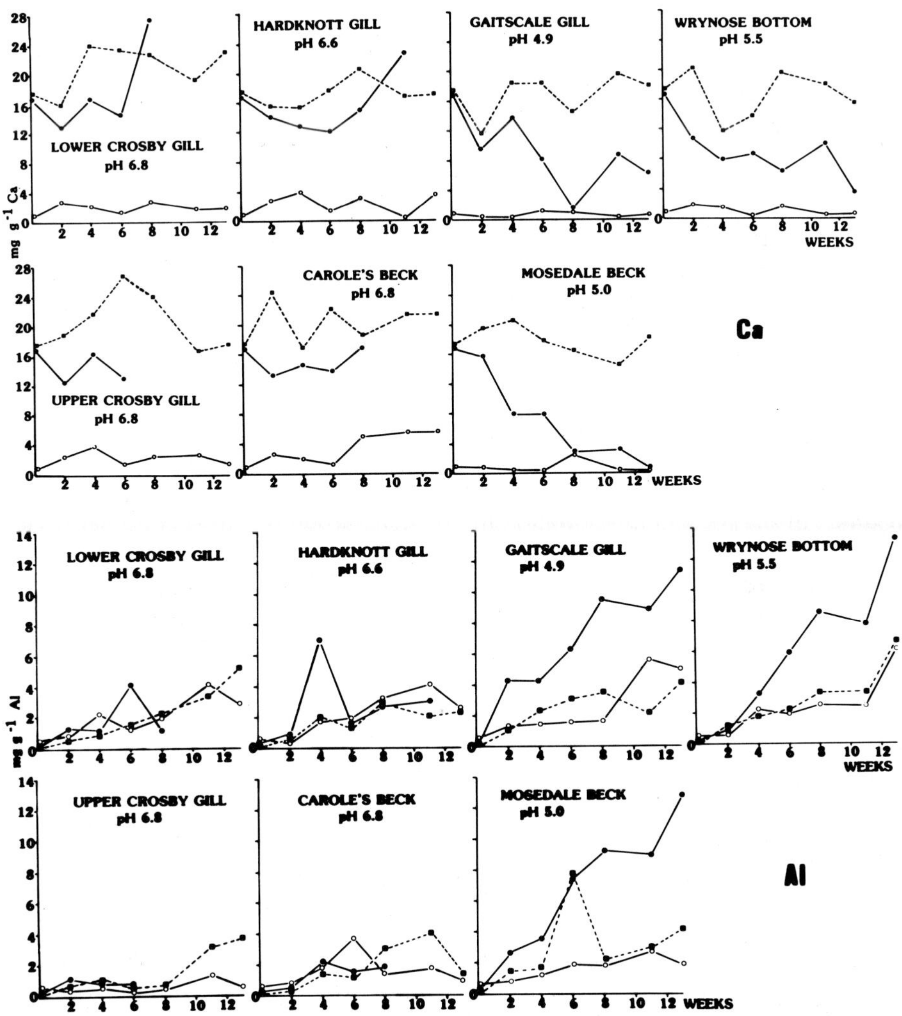

Fig. 8.2. Calcium and aluminum content of alder (●—●) and oak (■--■) leaves, and of mat grass (○—○) submersed in circumneutral and acid streams. Data from Chamier et al. (1989)

However, these workers made specific investigations of bacteria, not the fungi which may have been colonizing the leaves. Bacteria are known to be much less tolerant of low pH than fungi. Very low counts of bacteria colonizing leaves in acidic streams compared to circumneutral streams were made by Chamier (1987). Jones (1988), working on the same streams, found that bacterial genera colonizing leaves in circumneutral streams were absent from acidic waters.

Chamier et al. (1989) measured changes in the Na, K, Ca, Mg and Al content of decaying leaf litter submersed in the seven streams of the English Lake District described above (Table 8.1): three acidic and two circumneutral upland streams and two circumneutral lowland streams. The samples were taken from the same leaf packs that were analyzed for resident bacteria, aquatic hyphomycetes and weight loss.

Changes in Na, K and Mg content of alder, oak leaves and mat grass (*Nardus stricta* L.) were very similar in all streams, but changes in the Ca content of alder leaves were significantly greater in the acid streams all of which had very low calcium ion concentrations (see Table 8.1). In these, Ca levels of alder leaves dropped by 28% in the first two weeks of submersion and leaching of Ca continued for a further 11 weeks of sampling. Ca levels of alder leaves in the circumneutral streams remained fairly constant (Fig. 8.2). Over the same period, levels of Al rose consistently on alder leaves and after 8 weeks were 3.5 times higher in streams with a pH < 5.5 than in circumneutral streams (pH 6.6–6.8). The Al content of oak leaves and mat grass after 8 weeks' submersion was 1.2 times higher in acid than in circumneutral streams (see Fig. 8.2).

Aquatic hyphomycete colonization of alder leaves showed a consistent decline after six weeks' submersion in acid streams, associated with degradation rates four times slower than those in circumneutral streams (Fig. 8.3). On oak leaves with a higher, less labile calcium content and lower aluminum levels, aquatic hyphomycete colonization declined initially, but increased overall from 8–18 weeks in acid streams where decay rates were half those of circumneutral streams (Fig. 8.3).

Groom and Hildrew (1989) incubated a set each of bags of alder leaves (*Alnus glutinosa* (L.) Gaertn.) for 14 days, and beech leaves (*Fagus sylvatica* L.) for 28 days in an acid stream (pH 3.8–5.2) and a circumneutral stream (pH 6.5–7.2) in Kent. They then exposed some of the bags from the acid stream in the circumneutral stream and *vice versa*. Initial microbial respiration rates were higher on leaves incubated in the circumneutral than in the acid stream. After 32 days, respiration rates on leaves incubated in the acid and exposed in the circumneutral stream were equal to rates on leaves permanently in circumneutral conditions. Conversely, respiration on leaves incubated in circumneutral conditions and transferred to acid water was not significantly different from respiration on leaves permanently in acidic conditions. Differences in respiration were much more marked in alder, which degrades quickly, than on beech leaves which degrade slowly.

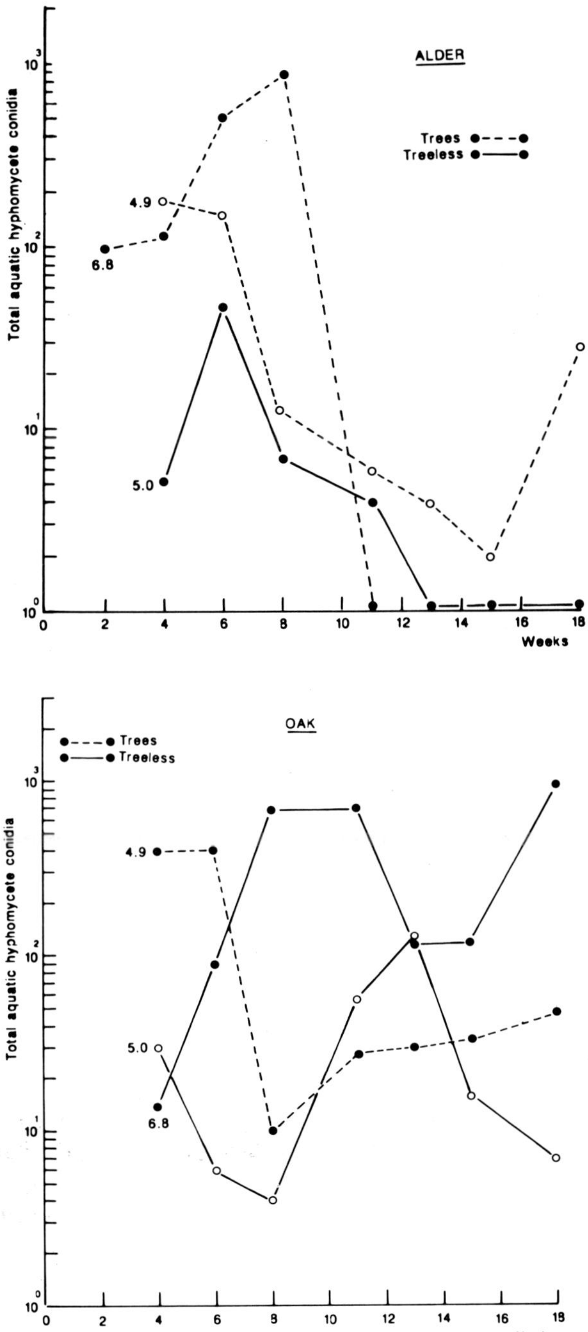

Fig. 8.3. Total conidial counts from standardized strips cut from *alder* and *oak* leaves submersed in circumneutral and acid streams. Data from Chamier (1987)

Direct counts of fungi and bacteria colonizing the leaves supported the evidence of respiration measurements. From alder incubated in the circumneutral stream, conidial counts were ten times higher (six species of aquatic hyphomycetes) than from leaves incubated in the acid stream (two species). Bacterial counts were three times higher on circumneutral than on acid conditioned alder. Similar results were obtained from beech leaves, which, like oak, degrade at about half the rate of alder, and attract much higher levels of aquatic hyphomycete colonization. These results are in line with those of Chamier (1987) reported above. Like Chamier et al. (1989), Groom and Hildrew (1989) found that although aluminum levels increased with time on leaves in both streams; they were highest on alder in acid streams. No measurements were made in this study of Ca levels in the streams or leaves.

These studies provide some evidence that where levels of calcium are low and aluminum levels rise above a threshold, the effect on aquatic hyphomycete metabolism is deleterious. Hall et al. (1985) reported that aluminum occupies adsorption sites on cell and tissue surfaces and may interfere with ion-exchange mechanisms. Lack of calcium ions may result in aluminum binding to membranes and proteins at sites normally occupied by calcium.

In experiments which seem appropriate to this topic, Abel and Bärlocher (1984) found that cadmium was more toxic to aquatic hyphomycetes in softwater with low calcium concentrations than in hardwater with high calcium levels. Using plate cultures they found that calcium ions gave protection against cadmium toxicity ($>0.1\,mg\,l^{-1}$) which affected the capacity of aquatic hyphomycetes both to grow and to sporulate. Abel and Bärlocher (1988) indicated from experimental data that accumulation of cadmium by aquatic hyphomycetes was due to surface adsorption and not to active uptake (see Chap. 9).

8.3.4 Studies in Circumneutral and Alkaline Streams

Studies of circumneutral (softwater) and alkaline (hardwater) streams by Bärlocher and Rosset (1981), Bärlocher (1982) and Wood-Eggenschwiler and Bärlocher (1983) indicate that though species richness of aquatic hyphomycetes is highest in circumneutral streams, metabolism (conidial production) is enhanced in alkaline streams. Softwater streams have much lower calcium ion concentrations (5.5–$31.6\,mg\,l^{-1}$) and alkalinity (0.17–$1.88\,meq\,l^{-1}$) than hardwater streams (Ca^{2+} 60.6–$81.7\,mg\,l^{-1}$; alkalinity 3.33–$4.50\,meq\,l^{-1}$) (see Table 8.1). Whilst species richness could be related to pH, conidial concentrations could not, but related best to the species richness of the vegetation. However, aquatic hyphomycete isolates from softwater streams survived as well as hardwater species in either type of stream (Rosset and Bärlocher 1985). Chamier and Dixon (1983) found that

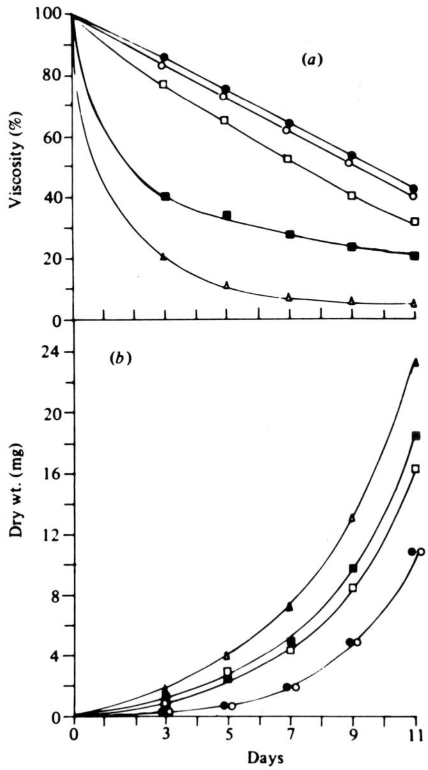

Fig. 8.4. a Viscosity loss of sodium polypectate medium inoculated with *Tetrachaetum elegans*. Average of 3 replicates. Initial pH = 6.8; final pH = 7.8; **b** growth curves of *T. elegans* from the same culture. Average of 3 replicates. Initial Ca^{2+} concentrations in mg l^{-1}: ○, 0; ●, 10; □, 20; ■, 40; ▲, 80. Data from Chamier and Dixon (1983)

increasing calcium ion concentration at a pH > 6.5 stimulated production of pectin lyase by aquatic hyphomycetes which increased the rate of leaf maceration (Fig. 8.4). This, combined with a wider range of natural resources may explain the enhanced metabolism of aquatic hyphomycetes in hardwater streams. Species richness is not a reliable guide to physiological performance. At a pH of 6.8, Chamier (1987) found that two species of aquatic hyphomycetes in a treeless moorland stream degraded alder leaves four times more quickly than five species in an acid stream pH 4.9 (see Table 8.2; Fig. 8.1).

8.3.5 Laboratory Studies of pH Tolerances

In the context of field studies, what is to be made of the conflicting results from laboratory studies of the effect of pH on aquatic hyphomycetes?

They may be as misleading as trying to establish the natural history of the elephant from seeing one sitting up on a drum in a circus eating currant buns.

The first requirement of laboratory studies of aquatic hyphomycetes is that physico-chemical conditions should be as similar as possible to those observed and measured in the field. The ambient temperature is an important first consideration – optimal temperatures for growth and sporulation of aquatic hyphomycetes (15–20 °C: Thornton 1963; Koske and Duncan 1974) are well above average stream temperatures over the seasons when the fungi are most active in temperate latitudes. Yet laboratory investigations of pH optima of selected species have been conducted at 10, 15, 17 and 22 °C. Differences in temperature optima between species could be confused with pH preferences and investigations of individual species ignore the possibility of interactions (Webster et al. 1976).

The second consideration of importance in investigations of pH optima of aquatic hyphomycete species is that the water chemistry characteristic of acidic, circumneutral and alkaline streams should be followed. The pH of streamwater, as has been noted above, is not a factor independent of other chemical parameters. Yet all experiments of pH optima have been based on general synthetic salts media adjusted with HCl or NaOH and frequently buffered with phosphate. Pristine freshwater streams have very low to undetectable levels of phosphate (Chamier 1980; Wood-Eggenschwiler and Bärlocher 1983; Carrick and Sutcliffe 1983), and many aquatic hyphomycete species are extremely sensitive even to low levels of phosphate in a medium (Chamier, unpubl. obs.). Iqbal (1972) found that the concentration of phosphate buffer affected the ability of aquatic hyphomycete species he tested to grow and sporulate at different pH values. *Articulospora tetracladia*, for example, showed poor tolerance of high phosphate concentrations above pH 6. But at low buffer molarity, pH tolerance went up to 8. The effects of calcium ion concentration on growth of aquatic hyphomycetes has been measured using phosphate buffers (Rosset and Bärlocher 1985). This leads to the formation of calcium phosophate which alters the original concentration on calcium ions in the medium. Actual free ion concentrations need to be measured over the course of such experiments.

The third requirement of laboratory experiments testing growth, sporulation, pH optimum or indeed any physiological phenomenon, is that the growth substrate (carbon source) should be either a natural substrate, or a model analogue. Aquatic hyphomycetes do not grow on malt agar in streams.

No laboratory investigations of pH optima and tolerances of aquatic hyphomycetes meet all these requirements, which may account for the fact that results are contradictory and generally fail to match field observations. Solid media with fixed pH values are preferable to liquid cultures in flasks where, even in buffered solutions, fungal metabolism causes large changes in the pH of the medium (Chamier 1980, Thompson and Bärlocher 1989).

Results of pH optima (4–5) of ten species of aquatic hyphomycetes by Rosset and Bärlocher (1985) were made on phosphate-buffered media with malt extract as a carbon source at 10 °C. Experiments by Thompson and Bärlocher (1989) on three species used synthetic salts medium, apparently without a nitrogen or phosphate source, adjusted with HCl or NaOH at 22 °C with autoclaved leaves as a substrate. The results were inconclusive.

Chamier and Dixon (1982b) and Chamier et al. (1984), investigating pectinase activity in nine aquatic hyphomycete species on solid medium at 15 °C with sodium polypectate (an analogue of a natural substrate) as a carbon source at fixed pH 5 or pH 7, found that although all species could elaborate polygalacturonases at pH 5 as a constitutive enzyme, only five could grow; at pH 7 all species could elaborate pectin lysases and all grew. These experiments were well replicated and agree closely with field observations of the species investigated. The species that grew at pH 5 and 7 have been found in acid and circumneutral to alkaline streams: *Varicosporium elodeae*, *V. giganteum*, *Articulospora tetracladia*, *Tricladium splendens*. Species that grew only at pH 7 were present only in circumneutral to alkaline streams or infrequent in acid streams: *Mycocentrospora angulata*, *Tetrachaetum elegans* and *Lemonniera aquatica* (Iqbal and Webster 1973, 1977; Shearer and Webster 1985).

8.3.6 Laboratory Studies of pH and Activity of Cell-Wall Degrading Enzymes

Cell-wall polymers of plants appear to be the natural substrates for aquatic hyphomycetes. Most of the enzymes which degrade these polymers are reported to have acidic pH optima (Table 8.3; based on published data). However, when individual species of aquatic hyphomycetes have been investigated for polysaccharidase capability, it has been shown that they vary

Table 8.3. Plant cell-wall polymerases and their pH optima

Polymer	Depolymerizing enzyme	Optimum
Pectic polymers	Polygalacturonase	5.0–5.6
	Pectin lyase	8.0–9.0
Hemicelluloses		
L-arabinan	L-arabinanase	2.5–6.0
D-galactan	D-galactanase	6.0–7.0
D-mannan	D-mannanase	6.0
D-xylan	D-xylanase	3.5–5.5
Cellulose	Exo-glucanase	
	Endo-glucanase	c. 5.0
	Cellobiase	
Lignin	Ligninase	Not known

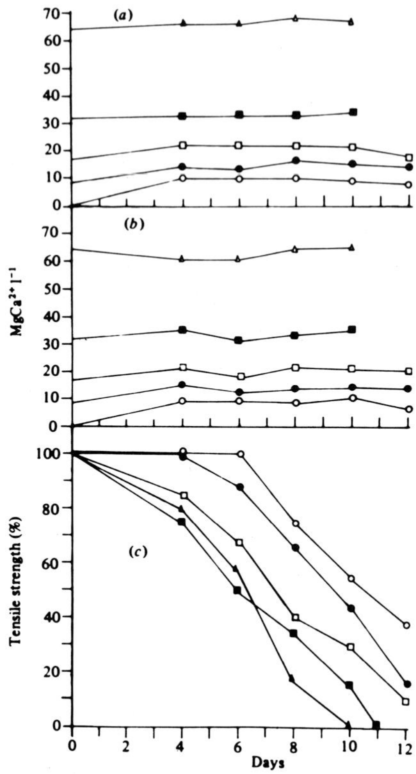

Fig. 8.5. **a** Ca^{2+} in $mg\,l^{-1}$ of supernatant of control alder leaf strips (without inoculum); **b** Ca^{2+} of supernatant with inoculated strips; **c** loss of tensile strength of alder leaf strips inoculated with *Tetrachaetum elegans*. Average of 3 replicates, as % of control strips. Initial pH = 6.7; final pH = 7.1. Initial Ca^{2+} concentrations in $mg\,l^{-1}$: ○, 0; ●, 10; □, 20; ■, 40; ▲, 80. Data from Chamier and Dixon (1983)

considerably. Optimal conditions may be less important than range. For example, polygalacturonases of *Tricladium splendens* have pH optima from 5–5.6, but show activity between pH 4–7. Similarly, pectinlyases of *Tetrachaetum elegans* and *Mycocentrospora angulata* with a pH optimum of 9, are active from 6.5–10, and have been shown to be involved in leaf maceration at pH 7. Theoretically, therefore, pectic polysaccharides, which constitute about 34% of the weight of primary plant cell-walls, could be degraded by aquatic hyphomycete pectinases from pH 4–10 (Chamier and Dixon 1982b). At a pH > 6.5, increasing calcium ion concentration would increase pectin lyase production and enhance enzyme activity, leaf maceration and with it growth and metabolism of the fungus (Chamier and Dixon 1983; Figs. 8.4, 8.5). Moreover, species may be physiologically fine-tuned to the level of calcium ions in their stream. Growth and pectin lyase production by *Tetrachaetum elegans* at pH 7 on a pectic carbon source was compared using distilled water, or filtered stream water, or 50% distilled water/50% mineral salts solution, or distilled water with 0.5 mM $CaCl_2$, or distilled water with 0.1 mM NaCl. The fungus grew best in the streamwater medium and its $CaCl_2$ equivalent (Chamier 1980).

Pectic polymers, constituting such a large component of non-lignified plant cell-walls and being the most accessible of the polymers, must be an important substrate for aquatic hyphomycetes. None of the 15 species investigated is without a pectinase or group of pectinases.

Hemicelluloses constitute about 24% of plant cell-wall polymers. Reported pH optima range from 2.5–7.0. Investigation of hemicellulase production by aquatic hyphomycetes has been limited to xylanases (Suberkropp and Klug 1980; Chamier et al. 1984). Three out of five species showed xylanase activity at pH 7 and 5; four out of six species at pH 6.2.

Cellulose, constituting about 20% of the weight of plant cell-wall polymers, is the least accessible. Ability of degrade native cellulose at pH 5 (the reported optimum for cellulases) was found in only half of 14 species of aquatic hyphomycetes tested (Chamier 1985). Three out of six species grew on microcrystalline cellulose as the sole carbon source at pH 6.2. The isolates were all from the same stream with an average pH of 7. The pH range of cellulase activity for enzymes secreted by aquatic hyphomycetes has not been established.

On the basis of this slender evidence, it seems that neutral/alkaline streams would provide an optimal pH environment for pectinase activity in all aquatic hyphomycetes colonizing leaves which may help to explain why conidial concentrations are highest in these streams. Circumneutral streams would be sub-optimal for pectinases, but would include activity of hemicellulases and cellulases in species which elaborate them. These conditions would favour the high species diversity found in these waters. In theory, all polymerases investigated should be active in acidic streams. Low species numbers and slow rates of decomposition in acidic streams support the thesis that factors of the water chemistry inhibit fungal metabolism.

8.4 Nitrogen and Phosphorus

In addition to pH and its associated water chemistry, levels of dissolved nitrogen and phosphorus in streams are biologically important. These are nutrients for which aquatic fungi would have to compete with other microbes and plants. Field and laboratory experiments on rates of decay of leaves report that these increase with nitrogen and phosphorus levels in the water (Kaushik and Hynes 1971; Howarth and Fisher 1976; Meyer 1980; Meyer and Johnson 1983). Studies by Triska et al. (1975) and Elwood et al. (1981) suggest that nitrogen and phosphorus levels are an important limiting factor in rates of detrital processing in streams. Workers concluded that increased microbial metabolism on leaf substrates is responsible for the

Table 8.4. Nitrate and phosphate levels in streams related to decay rates of alder leaves and average number of species of aquatic hyphomycetes

Stream	pH	NO_3-N $mg\,l^{-1}$	PO_4-P $\mu g\,l^{-1}$	Weeks to complete disappearance	Number of species	Reference
River Bourne	7.0	0.35–2.5	0–80	11	23	Chamier and Dixon (1982a)
Four Lake District streams	6.8	0.04	0–50	11	13	Chamier (1987)
Dartmoor streams	5.0	0.04	?	?	12	Iqbal and Webster (1977)
River Exe and tributaries	7.5	0.27–4.4	?	?	25	Iqbal and Webster (1973)

effects observed, but none distinguished between bacteria and fungi or specific aquatic hyphomycetes. In laboratory studies, Thornton (1963) found that the aquatic hyphomycetes he tested were able to use inorganic nitrate and ammonium ions, a feature uncommon in other aquatic fungi.

Reports on the water chemistry of streams usually do not include nitrogen or phosphorus levels – particularly the latter. Much depends on whether the stream has a catchment area under fertilized cultivation or is polluted with waste. NO_3-N in the River Bourne, Surrey, (Chamier and Dixon 1982a) varied from 0.35–2.5 $mg\,l^{-1}$ over a 6-month period from autumn to spring, with samples taken every 2 weeks. PO_4-P varied from undetectable levels to 80 $\mu g\,l^{-1}$ over the same period. Regardless of pH, NO_3-N values for seven streams of the English Lake District sampled at 2-week interval at the same time of year were about 0.04 $mg\,l^{-1}$ and PO_4-P values from 0–50 $\mu g\,l^{-1}$ (Carrick and Sutcliffe 1983; Chamier 1987). Though nitrate levels were ten times higher in the Surrey stream (phosphate strictly comparable), total decay time for alder leaves were very similar to those in the Lake District streams of the same pH (6.8–7) – 11 weeks on average (Table 8.4).

Nitrate levels in the River Exe and its tributaries (pH 6.3–8.7) which have a rich flora of aquatic hyphomycetes (Iqbal and Webster 1973) range from 0.27–4.4 $mg\,l^{-1}$, similar to the River Bourne. Dartmoor streams (pH 4–6.45) with an impoverished mycoflora have NO_3-N values similar to the Lake District streams (Table 8.4). NH_4-N concentrations in Lake District streams are barely detectable at any pH (Carrick and Sutcliffe 1983) and range from 30–200 $\mu g\,l^{-1}$ in the River Exe and its tributaries (Iqbal and Webster 1973).

Suberkropp (1991) investigated changes in microbial biomass, respiration and sporulation rates of aquatic hyphomycetes colonizing tulip-poplar leaf disks placed in two streams of contrasting water chemistry. South Sandy Creek (pH 6.2) had an average alkalinity of 0.08 $meq\,l^{-1}$, NO_3-N of 6.4 $\mu g\,l^{-1}$ and PO_4-P of 19.9 $\mu g\,l^{-1}$. Mud Creek (pH 8.3) had significantly higher values: average alkalinity was 2.9, NO_3-N was 305.3, and PO_4-P was 86.4. Stream temperatures over the experimental period were comparable.

Leaf disks were degraded faster in Mud Creek than in South Sandy Creek, and microbial activity was higher. Decay coefficients were -0.0172 (Mud Creek) and -0.0115 day^{-1} (South Sandy Creek). Maximum ATP concentrations of leaf disks were 4.1 times, and respiration rates 2.9 times higher in the Mud Creek. Conidium production from Mud Creek disks exceeded production from South Sandy Creek by a factor of 7.4.

This study indicates that the effects of nitrogen and phosphate levels on aquatic hyphomycete metabolism and communities, which may modify the influence of alkalinity and pH, cannot be ignored.

It is impossible to generalize from this evidence, but the nitrogen levels vary considerably in the streams studied and may influence the number of species of aquatic hyphomycetes in a stream and their metabolism. The sensitivity of aquatic hyphomycetes to phosphate would appear to be an adaptation to the very low levels available in freshwater streams.

8.5 Species Composition and Water Chemistry

The relationship between species composition of the aquatic hyphomycete community in a stream or river and its water chemistry is difficult to establish. Very few species lists are published with relevant data on water chemistry obtained through frequent measurements over a sustained period of time. Wood-Eggenschwiler and Bärlocher (1983) found that streams in the same region with similar water chemistry had a higher percentage of species in common than streams of contrasting water chemistry in a different region. Chamier (1980), in a reciprocal averaging ordination of 59 published species lists from around the world, found that the nearest neighbour to the River Bourne (Surrey) in species composition was the River Creedy (Devon). Their water chemistry was very similar, though geographically they are distinct.

8.6 Further Research

Given the fragmentary nature of research into the relationship between water chemistry and aquatic hyphomycetes, there are many avenues that could be pursued profitably. To be significant, field studies should always include comprehensive physico-chemical data of the water body. The weaknesses of many field studies of aquatic hyphomycetes are: infrequent sampling of all the factors under investigation; absence of replication and the tendency to compile species lists which add nothing to the understanding of fungal physiology. Some of the methods employed in field studies are inadequate if used alone. Foam samples may give a reasonably full account of species diversity but are unreliable for assessing the relative abundance of species. Random leaf or wood samples give no clues as to how long the material has been submersed in the water so that the stage of fungal colonization is unknown – the number of species and the extent of colonization range along a cycle from initial submersion to complete degradation (Chamier and Dixon 1982a; Chamier et al. 1984). Filtration of conidia onto millipore filters which are then stained and examined microscopically, though frequently employed, is a difficult method for identifying all but the obvious species. Subsamples from incubated material, sedimented down and viewed through an inverted microscope, give a better image of conidia for identification and allow for comparative quantification. Material from the original sample can be grown in pure culture for reliable identification of common species. Identifications based purely on conidial form are open to error. Conidial concentrations in stream water need to be based on frequent sampling, and conidial counts made from incubated litter should be taken from standardized areas or weights of material.

The subject lends itself to laboratory experimentation, but, as has been discussed above, this needs to be conducted strictly under conditions derived from field studies. This includes the use of appropriate substrates as a carbon source and a demonstration that the conidia under investigation can germinate and grow in the conditions provided. Of particular interest is the effect of soluble, monomeric aluminum on fungal metabolism at low pH values and the relationship between aluminum and calcium ions. Investigations of the effects of inorganic nitrogen and phosphorus levels on the metabolism of aquatic hyphomycetes would aid understanding of the effects of nutrient enrichment in streams. It needs to be emphasized that what happens to microbes degrading allochthonous litter in running waters may have consequences for other trophic levels in the ecosystem.

References

Abel TH, Bärlocher F (1984) Effects of cadmium on aquatic hyphomycetes. Appl Environ Microbiol 48:245–251

Abel TH, Bärlocher F (1988) Uptake of cacmium by *Gammarus fossarum* (Amphipoda) from food and water. J Appl Ecol 25:223–231

Bärlocher F (1982) Conidium production from leaves and needles in four streams. Can J Bot 60:1487–1494

Bärlocher F (1987) Aquatic hyphomycete spora in 10 streams of New Brunswick and Nova Scotia. Can J Bot 65:76–79

Bärlocher F, Petersen RC (1988) Aquatic hyphomycetes in an acidified and limed humic stream in Sweden. Verh Int Ver Limnol 23:1279–1286

Bärlocher F, Rosset J (1981) Aquatic hyphomycete spora of two Black Forest and two Swiss Jura streams. Trans Br Mycol Soc 76:479–483

Carrick TR, Sutcliffe DW (1983) Concentrations of major ions in streams on catchments of the River Duddon (1971–1974) and Windermere (1975–1978), English Lake District. Freshwater Biol Assoc Occ Publ No 21

Chamier A-C (1980) Pectinases in leaf degradation by aquatic hyphomycetes. PhD Thesis, University of London

Chamier A-C (1985) Cell-wall degrading enzymes of aquatic hyphomycetes: a review. Bot J Linn Soc 91:67–81

Chamier A-C (1987) Effect of pH on microbial degradation of leaf litter in seven streams of the English Lake District. Oecologia 71:491–500

Chamier A-C, Dixon PA (1982a) Pectinases in leaf degradation by aquatic hyphomycetes. I. The field study: the colonisation-pattern of aquatic hyphomycetes on leaf packs in a Surrey stream. Oecologia 52:109–115

Chamier A-C, Dixon PA (1982b) Pectinases in leaf degradation by aquatic hyphomycetes: the enzymes and leaf maceration. J Gen Microbiol 128:2469–2483

Chamier A-C, Dixon PA (1983) Effect of calcium-ion concentration on leaf maceration by *Tetrachaetum elegans*. Trans Br Mycol Soc 81:415–417

Chamier A-C, Dixon PA, Archer SA (1984) The spatial distribution of fungi on decomposing alder leaves in a freshwater stream. Oecologia 64:92–103

Chamier A-C, Sutcliffe DW, Lishman JP (1989) Changes in Na, K, Ca, Mg and Al content of sumbersed leaf litter, related to ingestion by the amphipod *Gammarus pulex* (L.). Freshwater Biol 21:181–189

Elwood JW, Newbold JD, Trimble AF, Stark RW (1981) The limiting role of phosphorus in a woodland stream ecosystem: effects of P enrichment on leaf decomposition and primary producers. Ecology 62:146–158

Friberg F, Otto C, Svensson B (1980) Effects of acidification on the dynamics of allochthonous leaf material and benthic invertebrate communities in running water. In: Drablos D, Tollan A (eds) Ecological Impact of Acid Precipitation. Proc Int Conf Oslo Norway. SNSF Proj 304–305

Groom AP, Hildrew AG (1989) Food quality for detritivores in streams of contrasting pH. J Anim Ecol 58:863–881

Hall RJ, Driscoll CT, Likens GE, Pratt JM (1985) Physical, chemical and biological consequences of episodic aluminium additions to a stream. Limnol Oceanogr 30:212–220

Howarth RW, Fisher SG (1976) Carbon, nitrogen and phosphorus dynamics during leaf decay in nutrient enriched stream micro-ecosystems. Freshwater Biol 6:221–228

Hynes HBN (1970) The ecology of running waters. University of Toronto Press, Toronto

Hynes HBN (1983) Groundwater and stream ecology. Hydrobiologia 100:93–99

Iqbal SH (1972) Some observations on aquatic hyphomycetes. PhD Thesis, University of Exeter, Devon

Iqbal SH, Webster J (1973) Aquatic hyphomycete spora of the River Exe and its tributaries. Trans Br Mycol Soc 61:331–346

Iqbal SH, Webster J (1977) Aquatic hyphomycete spora of some Dartmoor streams. Trans Br Mycol Soc 69:233–241

Jones JG (1986) Bacteria in soft and acid streams. In: Sutcliffe DW (ed) Acidification problems of freshwaters: trophic relationships. Freshwater Biol Assoc Rep, pp 29–31

Kaushik NK, Hynes HBN (1971) The fate of the dead leaves that fall into streams. Arch Hydrobiol 68:1465–1515

Kerekes JJ (1989) Acidification of organic waters in Kejimkujik National Park, Nova Scotia. Kluwer, Dordrecht

Koske RE, Duncan IW (1974) Temperature effects on growth, sporulation and germination of some aquatic hyphomycetes. Can J Bot 52:1387–1391

Mackay RJ, Kersey KE (1985) A preliminary study of aquatic insect communities and leaf decomposition in acid streams near Dorset, Ontario. Hydrobiologia 122:3–11

Marvanová L (1984) Conidia in water on the protected area Slovensky Raj. Biologia (Bratislava) 39:821–832

Meyer JL (1980) Dynamics of phosphorus and organic matter during leaf decomposition in a freshwater stream. Oikos 34:44–53

Meyer JL, Johnson C (1983) The influence of elevated nitrate concentration on rate of leaf decomposition in a stream. Freshwater Biol 13:177–183

Mulholland PJ, Palumbo AV, Elwood JW (1987) Effects of acidification on leaf decomposition in streams. J N Am Benthol Soc 6:147–158

Palumbo AV, Mulholland PJ, Elwood JW (1987) Micobial communities on leaf material protected from macroinvertebrate grazing in acidic and circumneutral streams. Can J Fish Aquat Sci 44:1064–1070

Rosset J, Bärlocher F (1985) Transplant experiments with aquatic hyphomycetes. Verh Int Ver Limnol 22:2786–2790
Regelsberger B, Messner K, Descals E (1987) Species diversity of aquatic hyphomycetes in four Austrian streams. Mycotaxon 30:439–454
Shearer CA, Webster J (1985) Aquatic hyphomycete communities in the River Teign. I. Longitudinal distribution patterns. Trans Br Mycol Soc 84:489–501
Suberkropp K (1991) Relationships between growth and sporulation of aquatic hyphomycetes on decomposing leaf litter. Mycol Res 95:843–850
Suberkropp K, Klug MJ (1980) The maceration of deciduous leaf litter by aquatic hyphomycetes. Can J Bot 58:1025–1031
Sutcliffe DW (1983) Acid precipitation and its effects on aquatic systems in the English Lake District (Cumbria). Freshwater Biol Assoc Rep No 51, pp 30–62
Sutcliffe DW, Carrick TR (1973) Studies on mountain streams in the English Lake district I. pH, calcium and the distribution of invertebrates in the River Duddon. Freshwater Biol 3:437–462
Suzuki S, Nimura H (1960a) Aquatic hyphomycetes in the lakes of Mt. Hakkoda. J Jpn Bot 35:265–268
Suzuki S, Nimura H (1960b) Ecological study on aquatic hyphomycetes in the Goshikinuma and Akanuma Lake group. Bot Mag Tokyo 73:360–364
Suzuki S, Nimura H (1961) Relation between the distribution of aquatic hyphomycetes in Japanese lakes and lake types. Bot Mag Tokyo 74:51–55
Thompson PL, Bärlocher F (1989) Effect of pH on leaf breakdown in streams and in the laboratory. J N Am Benthol Soc 8:203–210
Thornton DR (1963) The physiology and nutrition of some aquatic hyphomycetes. J Gen Microbiol 33:23–31
Tipping E (1990) A model of surface water acidification in Cumbria and its uses in long-term research. Freshwater Biol 23:7–23
Triska FJ, Sedell JR, Buckley B (1975) The processing of conifer and hardwood leaves in two coniferous forest streams. II. Biochemical and nutrient changes. Verh Int Ver Limnol 19:1628–1639
Webster J, Moran ST, Davey RA (1976) Growth and sporulation of *Tricladium chaetocladium* and *Lunulospora curvula* in relation to temperature. Trans Br Mycol Soc 67:491–549
Wood-Eggenschwiler S, Bärlocher F (1983) Aquatic hyphomycetes in sixteen streams in France, Germany and Switzerland. Trans Br Mycol Soc 81:371–379

9 Human Interference

Felix Bärlocher

9.1 Introduction

Ingold (1942, 1975) and other early students of aquatic hyphomycetes (e.g. Petersen 1962, 1963a,b, Nilsson 1964) quickly established that these fungi are most common and most diverse in clean, well-oxygenated streams running through forests. Human activities will modify these conditions. Some of the changes may be beneficial to the fungi; more realistically, one would expect a usually deleterious balance between positive and negative effects. Thus, removing a few riparian trees will lower the food base of the fungi and of leaf-shredding invertebrates. It is conceivable that, on balance, the fungi might gain more by being less exposed to the invertebrates than by losing substrata. But drastic reductions in the riparian vegetation result in a clear impoverishment of the fungal community (Metvalli and Shearer 1989; Chergui 1990; Chap. 3). It is due to the dilution of available food resources. A similar effects occurs naturally when the river widens (Nilsson 1964; Chauvet 1989), since leaf input is related primarily to the length of the river edge, while water volume increases with river width (Bird and Kaushik 1981).

Two side-effects of partial or complete removal of riparian vegetation are increased water temperature (which may favour species such as *Lunulospora curvula*, Suberkropp 1984) and erosion (Hynes 1970). Increased sediment load of stream water may alternately sandblast or smother leaves and their associated fungal community.

When the course of a river is straightened, or blockages are removed, to improve drainage of the catchment area, the channel becomes less retentive. As discussed in Chapter 2, downstream displacement of both leaves and fungal spores will increase, which will lower the food supply as well as the fungi's ability to capture it.

In addition to changing watercourses and the surrounding vegetation, human activities increase the amounts of many chemicals introduced into running waters.

9.2 Organic Pollution

The apparent preference of aquatic hyphomycetes for fast-running, well-aerated streams indicates that they cannot tolerate low oxygen levels. This is probably true: they are much less common on leaves turned black in the absence of oxygen (unpubl. obs.). Nevertheless, of four species tested, three survived up to 12 months under anaerobic conditions, whereas one failed to survive even 3 months (Field and Webster 1983). For all four, percentage survival was lower than for seven aero-aquatic fungi. Since a major consequence of organic pollution is increased oxygen demand, one might expect that it will be detrimental to aquatic hyphomycetes. On the other hand, aquatic hyphomycetes can obviously use dissolved organic matter as nutrients (allowing their growth in broth cultures; they also make use of leaf leachates, Bärlocher and Kendrick 1974). Not surprisingly, studies investigating the effects of organic pollution have given conflicting results.

In two surveys, both Greathead (1961) and Conway (1970) reported greatly reduced numbers of aquatic hyphomycetes in stream sites contaminated by sewage effluents. Burgos and Castillo (1986) compared the colonization of willow leaves in a stream above and below a source of organically polluted water. All leaves in the clean zone were colonized and a total of 17 species were found; in the polluted zone, only 2.5% of the leaves were colonized by a total of four species. As a result of the pollution, oxygen saturation dropped to 15%. Au et al. (1990) found 27 aquatic species in a clean forest stream in Hong Kong; in another stream, polluted by animal wastes and with an oxygen deficit of 85%, this number dropped to 14. Terrestrial fungi were more diverse in the polluted stream.

By contrast, Sládecková (1963) reported that aquatic hyphomycetes appeared to be stimulated by starch factory wastes. However, in both years of her study, the factory apparently became active in October. Aquatic hyphomycetes appeared at about the same time. It is possible that the timing is simply due to the normal seasonal fluctuations of the aquatic hyphomycetes (Chap. 3). In addition, counts of the fungi were based on conidia settling on solid substrata made of wood or plastic. They might have been swept in from stream sections above the starch factory.

Van der Merwe and Jooste (1988) compared the fungal communities at two sites of the Mooi River in South Africa. Again, one site was unpolluted, the other was downstream of a sewage processing plant. In summer, the oxygen content of the polluted site was 30–60% (average 37; 100% = oxygen content of clean site); in winter, 33–100% (average 72). In summer, seven species declined at the polluted site, two increased and one remained the same. In winter, the differences were less pronounced. In all instances, the leaves were decomposed in approximately eight weeks.

In a comparable study in Germany, Suberkropp et al. (1988) investigated the effects of sewage treatment plant effluents. Phosphate, ammonia, nitrite, dissolved organic matter, BOD_5, and bacteria all increased below the plant.

The oxygen content dropped from 106 to 95% of the theoretical saturation value. The community structure of aquatic hyphomycetes, characterized by water filtration and study of randomly collected as well as introduced leaves, remained unchanged.

In several studies related to this topic, Bengtsson (1982, 1983, 1988) studied the behaviour of selected fungal species when amino acids were added to leaves decomposing in stream or tap water. This resulted in shifts of dominance patterns, and, at higher concentrations, it delayed breakdown of macromolecules such as albumin and cellulose. In all cases, the medium was aerated to ensure an adequate supply of oxygen.

I think one has to conclude that the addition of organic pollutants may change the abundances of individual fungal species and delay the breakdown of leaves and other plant detritus. It will not necessarily lower overall activity of the aquatic hyphomycete community. This becomes more likely when it leads to a substantial reduction in the oxygen content of the stream water.

9.3 Heavy Metals

Human activities have greatly increased the mobility of heavy metals in the biosphere. They may enter streams by industrial and domestic effluents or by precipitation. Little is known about their effects on aquatic fungi.

Duddridge and Wainwright (1980) found that both Pb^{2+} and Cd^{2+} inhibited growth of three fungi (none of them an aquatic hyphomycete). Low concentrations of Cd^{2+} ($0.5–1\,mg\,l^{-1}$) stimulated mycelial growth; Pb^{2+} ions had no discernible effects at these concentrations. Zn^{2+} ions, on the other hand, stimulated growth or had no effect at $1–10\,mg\,l^{-1}$. All fungi accumulated considerable amounts of the three metals, generally in the order Cd > Pb > Zn. This was largely due to an adsorptive process involving organic ligands on the surface of the mycelium. Accumulation by heat-killed mycelium followed the same pattern, but did not reach the same levels, indicating that heat changes or destroys some metal-binding ligands. Mycelium of the three fungi supported growth by the amphipod *Gammarus pulex*, but survival was reduced when they were fed Cd contaminated mycelium. The dead bodies contained Cd, confirming that it can be transferred from aquatic fungi to invertebrates.

Appreciable amounts of Cd were accumulated by all three fungi when they were grown in several changes of a medium with a low concentration of the metal ($0.1\,mg\,l^{-1}$). This simulates sporadic exposure to low heavy metal concentrations.

The effects of Cd on aquatic hyphomycetes were studied by Abel and Bärlocher (1984). Sporulation was not affected at concentrations below $0.1\,mg\,l^{-1}$. LC_{50} (concentration at which spore production is reduced by 50%) was 0.4 in softwater and 12.1 in hardwater. Generally, *Heliscus lugdunensis* and *Tetracladium marchalianum* were the least sensitive species.

Growth on malt extract agar was also inhibited by Cd, though higher levels had to be added for a measurable effect ($>1\,\mathrm{mg\,l^{-1}}$). In general, there was a good correlation with the spore production data.

The addition of Ca and Zn ions to the growth medium reduced the toxic effect of Cd. A similar effect was found with EDTA when it was added at the same order of magnitude as the Cd ions; at higher levels, EDTA itself was toxic. The most likely explanation is that Ca and Zn act as antagonists of Cd, since sizes and chemical properties of the three ions are similar. The toxic effect of Cd is generally assumed to be caused by erroneous binding to proteins. For example, metallothionein proteins are known to irreversibly bind cadmium instead of zinc (Nakamura et al. 1981). Enzymes saturated with Cd lose their catalytic function. High Zn concentrations increase the chance of this ion, rather than Cd, being bound, which reduces the toxic effect. The protective effect of Ca is probably based on similar mechanisms.

In streams, natural organic ligands (e.g. humic acids, phenolics) might also interact with cadmium (Gjessing 1981). Similarly, EDTA has high complex formation constants with several cations. It is not a ligand specific for Cd, and forms complexes with ions essential for fungal growth. This might explains its toxicity towards fungi.

Abel and Bärlocher (1984) concluded that the lower sensitivity of fungi sporulating in hard water was almost certainly due to its higher Ca content. Calcium ions presumably compete with Cd for binding sites at enzymes or other proteins (e.g. carriers). Organic ligands were higher in the softwater, but apparently too low to have any effect at the Cd concentrations used.

Aquatic hyphomycetes in a liquid medium accumulated cadmium very rapidly and to high final concentrations (Abel and Bärlocher 1988), even at levels that interfered little with their growth and reproduction. Again, uptake patterns by live and dead mycelium suggested that passive adsorption to surface sites was the primary mechanism. Potentially, this biomagnification could be an important cause of mortality in leaf-eating invertebrates. Experiments with *Gammarus fossarum*, however, suggested that the direct effects of cadmium dissolved in the water are of greater significance. Leaves conditioned by fungi in water with a cadmium concentration of $1\,\mathrm{mg\,l^{-1}}$ reduced survival of the amphipods from 95 to 66% in 12 days; direct exposure to water with the same Cd concentration killed all animals in less than 10 days. Survival rates were higher in hard water, suggesting that Ca is able to displace Cd ions and thereby reduce its toxic effect on amphipods as well as on fungi. Abel and Bärlocher (1988) concluded that during persistent Cd pollution, uptake through water is more damaging than uptake through food, and that this difference is particularly pronounced in softwater. The situation might be reversed if the stream is subjected to a short pulse of pollution. Fungi immobilize cadmium very rapidly and efficiently. Stream invertebrates might escape the initial pulse, but suffer from ingesting cadmium-contaminated mycelium for an extended period of time. The relative insignificance of contaminated food becomes understandable when one compares the amount of water and food that pass through *Gammarus*. An animal weighing 10 mg requires 0.5 mg of food per day (Pieper 1978). In the same period, it has to filter 0.16 l of water to satisfy its oxygen demand. If the food consisted entirely of mycelium, the proportion between food and water would be 320 000 : 1. Even the high affinity between cadmium and fungal mycelium is insufficient to overcome these numbers.

9.4 Pesticides

Pentachlorophenol (PCP) is widely used in the treatment of both raw and processed lumber products, as an antifungal agent, an insecticide as well as a herbicide. PCP has long been known as a potent uncoupler of oxidative phosphorylation and inhibitor of various enzymes involved in glycolysis and the TCA cycle (Botstrom and Johansson 1972; Holmberg et al. 1972; Weinbach 1956). As a result, the metabolism of organisms faced with a PCP challenge becomes less efficient, and less energy will be available for growth and reproduction. To compensate, many organisms increase their basal respiration rate (Hodson and Blunt 1981).

When a pulse of PCP enters an aquatic system, it may be broken down by sunlight to less stable tetra- and trichlorophenols, whose phenolic rings can be cleaved by UV radiation and subsequently mineralized by macrophytes or microorganisms (Pignatello et al. 1983). Some PCP sinks to the light-poor benthic layer where it adheres to organic material such as detritus-rich silt or partly decayed leaves. Here, protected from exposure to light, it may persist at high concentrations for several years (Pierce and Viktor 1978).

PCP is highly lipophilic and tends to accumulate in organisms with high fat content. In decaying leaves, most of the lipids are concentrated in aquatic hyphomycetes (Cargill et al. 1985). Conidium production of these fungi was found to decline by a factor of approximately 20, when the PCP level in the water was increased from 0 to 10 mg l^{-1} (Bärlocher and Premdas 1988). Respiration peaked at 0.1 and 100 μg l^{-1}. By contrast, some terrestrial fungi show little or not sign of stress at PCP levels as high as 5000 mg l^{-1} (Cserjesi and Roff 1975).

As was the case with heavy metals, stream invertebrates might take up PCP through the water or with the food. The relative quantities involved make it likely that water is initially the more significant route, especially since PCP increases the respiration rate and therefore the uptake of water (Kierstead and Bärlocher 1989). However, the PCP level in the water column may drop to background levels within two months following a spill, whereas its concentration in the sediment can remain high for several years (Pierce and Viktor 1978).

Large scale production of DDT began in 1942 (Hodkinson 1976). It was banned in the early 1970s in the USA and many other countries, but is still widely used in Africa (Mpofu 1987). DDT has been detected in regions far removed from treated areas and is now considered to have a world-wide distribution. Its persistence and accumulation in fat tissues can produce serious effects in non-target organisms (Brown 1978). The first report on its effects on aquatic hyphomycetes was published by Dalton et al. (1970). They found that concentrations between 2 and $60\,mg\,l^{-1}$ increased the growth of four species in mineral medium enriched with yeast extract. Additional experiments indicated that DDT provides a source of carbon and energy which is most efficiently used when growth factors are provided as well (Hodkinson and Dalton 1973). On the other hand, when maltose, sucrose, lactose or arabinose were the main carbon source, DDT lowered the growth rate of *Heliscus submersus* (Hodkinson 1976). In these cases, the pH of the culture medium decreased due to the accumulation of pyruvate. This suggests that DDT interferes with acetyl Co-A formation. An increased respiratory quotient further indicated that DDT inhibits the normal aerobic oxidation of sugars and increases the fermentation process.

In vitro, several aquatic hyphomycetes can degrade DDT fairly rapidly. *Clavariopsis aquatica* converted 6% of the DDT in the medium to its metabolites, DDD and DDE, whereas *Heliscus submersus* converted 35% in 330 h (Hodkinson 1976). Why does DDT persist in nature? It has been suggested that its decomposition requires both anaerobiosis and subsequent aerobiosis (Pfaender and Alexander 1972).

Many of the rivers in the Western Ghat forests of India are lined by plantation crops. Pesticides used to protect the crops are carried into the streams by heavy monsoon rains and may affect aquatic hyphomycetes. Two herbicides (paraquat and 2,4-dichlorophenoxybutyric acid) and two fungicides (mancozeb and captafol) did not inhibit the growth of three aquatic hyphomycetes at concentrations below $10\,mg\,l^{-1}$ (Chandrashekar and Kaveriappa 1989). The two fungicides completely suppressed growth of all three species at a concentration $\geq 1\,g\,l^{-1}$; the two herbicides had a much weaker effect.

Some aero-aquatic fungi were strongly inhibited by the aquatic herbicide diquat bromide (Fronda and Kendrick 1985). On the other hand, *Pseudoaegerita matushimae* was almost unaffected. It did not appear to accumulate any diquat, either because it did not take it up or was able to detoxify it. The snail *Lymnaea* (Gastropoda) selectively ingested mycelium or leaves with the lowest accumulation of diquat (Fronda and Kendrick 1986). High concentration of the herbicide in the food lowered survival of the snail.

9.5 Acid Precipitation

The complex interactions between pH, alkalinity and dissolved inorganic ions are summarized in Chapter 8. It appears that low pH as such is well tolerated by many aquatic hyphomycetes. In fact, van Frankenhuyzen and

Geen (1986) found increased fungal growth on alder leaves when the pH was lowered from 6 to 4 by the addition of H_2SO_4, a treatment which simulates acid rain. In an artificial stream, weight loss of maple leaves due to microbial activity peaked at a pH between 5.5 and 6.0 (Thompson and Bärlocher 1989). In the same study, respiration and decompositional activities of three aquatic hyphomycetes did not change very drastically at pH levels between 4 and 9.

More significantly, lowered pH may increase biologically active aluminum or heavy metal ions. Their toxic effect will be more pronounced in poorly buffered (low calcium content) waters (Chap. 8). Since organic molecules often render dissolved metal ions non-toxic through complexation, it is important to distinguish between clear waters, where the pH has dropped due to acid precipitation, and brown water where the pH is low due to high concentrations of organic acids (Collier et al. 1990).

The toxicity of ionizable organic material is also pH dependent (Albert 1985). As the pH drops, many pesticides become more lipophilic and therefore will accumulate more readily in organisms (for example, DDT accumulation in fungi: Chacko and Lockwood 1967). It has been suggested that the same principle applies to natural organic substances. Due to acid rain, they will tend to become non-dissociated and more lipid soluble (Petersen and Persson 1987; Kullberg 1990). Again, this may increase their toxicity.

References

Abel TH, Bärlocher F (1984) Effects of cadmium on aquatic hyphomycetes. Appl Environ Microbiol 48:245–251

Abel TH, Bärlocher F (1988) Uptake of cadmium by *Gammarus fossarum* (Amphipoda) from food and water. J Appl Ecol 25:223–231

Albert A (1985) Selective toxicity. Chapman and Hall, New York

Au DWT, Hodkiss IJ, Vrijmoed LLP (1990) A study of the colonization of leaf packs in a non-polluted and a polluted stream in Hong Kong. In: Reisinger A, Bresinsky A (eds) Abstr 4th Int Mycological Congr, Regensburg, Botanical Institute, Univ Regensburg p 107

Bärlocher F, Kendrick B (1974) Dynamics of the fungal population on leaves in a stream. J Ecol 63:761–791

Bärlocher F, Premdas PD (1988) Effects of pentachlorophenol on aquatic hyphomycetes. Mycologia 80:135–137

Bengtsson G (1982) Patterns of amino acid utilization by aquatic hyphomycetes. Oecologia 55:355–363

Bengtsson G (1983) Habitat selection in two species of aquatic hyphomycetes. Microb Ecol 9:15–26

Bengtsson G (1988) The impact of dissolved amino acids on protein and cellulose degradation in stream waters. Hydrobiologia 164:97–102

Bird GA, Kaushik NK (1981) Coarse particulate organic matter in streams. In Lock MA, Williams DD (eds) Perspectives in running water ecology. Plenum, New York, pp 41–68

Botstrom SL, Johansson RG (1972) Effects of pentachlorophenol on enzymes involved in energy metabolism in the liver of the eel. Comp Biochem Physiol 41:359–369

Brown AWA (1978) Ecology of pesticides. Wiley and Sons, New York

Burgos EJ, Castillo PH (1986) Hyphomycetes acuaticos como indicadores de contaminacion. Biota 2:1–10
Cargill AS, Cummins KW, Hanson BJ, Lowry RR (1985) The role of lipids as feeding stimulants for shredding aquatic insects. Freshwater Biol 15:455–464
Chacko CI, Lockwood JL (1967) Accumulation of DDT and Dieldrin by microorganisms. Can J Microbiol 13:1123–1126
Chandrashekar KR, Kaveriappa KM (1989) Effect of pesticides on the growth of aquatic hyphomycetes. Toxicol Lett 48:311–315
Chauvet E (1989) Production, flux et décomposition des litières en milieu alluvial. Dynamique et rôle des hyphomycètes aquatiques dans le processus de décomposition. Thesis, Université de Toulouse, France
Chergui H (1990) The dynamics of aquatic hyphomycetes in an eastern Moroccon stream. Arch Hydrobiol 118:341–352
Collier KJ, Ball OJ, Graesser AK, Main MR, Winterbourn MJ (1990) Do organic and anthropogenic acidity have similar effects on aquatic fauna? Oikos 59:33–38
Conway KE (1970) The aquatic hyphomycetes of central New York. Mycologia 62:516–530
Cserjesi AJ, Roff JW (1975) Toxicity tests of some chemicals against certain wood-staining fungi. Int Biodeterior Bull 11:90–96
Dalton SA, Hodkinson M, Smith KA (1970) Interactions between DDT and river fungi. I. Effects of p,p′-DDT on the growth of aquatic hyphomycetes. Appl Microbiol 20:662–666
Duddridge JE, Wainwright M (1980) Heavy metal accumulation by aquatic fungi and reduction in viability of *Gammarus pulex* fed Cd^{2+} contaminated mycelium. Water Res 14:1605–1611
Field JI, Webster J (1983) Anaerobic survival of aquatic fungi. Trans Br Mycol Soc 81:365–369
Fronda A, Kendrick B (1985) Diquat uptake by four aero-aquatic hyphomycetes. Environ Pollut 37:229–244
Fronda A, Kendrick B (1986) Effects of food containing diquat on a pond invertebrate, *Lymnaea elodes*. Arch Hydrobiol 105:387–395
Gjessing ET (1981) The effect of aquatic humus on the biological availability of cadmium. Arch Hydrobiol 91:144–149
Greathead SK (1961) Some aquatic hyphomycetes in South Africa. J S Afr Bot 27:195–228
Hodkinson M (1976) Interactions between aquatic fungi and DDT. In: Jones EBG (ed) Recent advances in aquatic mycology. Paul Elek, London, pp 447–467
Hodkinson M, Dalton SA (1973) Interactions between DDT and river fungi. II. Influence of culture conditions on the compatability of fungi and p-p-DDT. Bull Environ Contam Toxicol 10:356–359
Hodson PV, Blunt BR (1981) Temperature-induced changes in pentachlorophenol chronic toxicity to early life stages of rainbow trout. Aquat Toxicol 1:113–127
Holmberg B, Jensen S, Larsson A, Lewander K, Olsson M (1972) Metabolic effects of technical pentachlorophenol (PCP) on the eel *Anguilla anguilla*. Comp Biochem Physiol 43:171–183
Hynes HBN (1970) The ecology of running waters. University of Toronto Press, Toronto
Ingold CT (1942) Aquatic hyphomycetes of decaying alder leaves. Trans Br Mycol Soc 25:339–417
Ingold CT (1975) An illustrated guide to aquatic and waterborne hyphomycetes (Fungi Imperfecti) with notes on their biology. Freshwater Biol Assoc Sci Publ No 30
Kierstead WG, Bärlocher F (1989) Ecological effects of pentachlorophenol on the brackish-water amphipod *Gammarus tigrinus*. Arch Hydrobiol 115:149–156
Kullberg A (1990) The octanol/water partition coefficient in absorptivity of DOC in 10 humic acid and neutral streams in southern Sweden. Verh Int Ver Limnol 24:169–172
Metvalli AA, Shearer CA (1989) Aquatic hyphomycete communities in clear-cut and wooded areas of an Illinois stream. Trans Ill Acad Sci 82:5–16
Mpofu SM (1987) DDT and its use in Zimbabwe. Zimbabwe Sci News 21:32–36

Nakamura Y, Katayama S, Okada Y, Suzuki F, Nagata Y (1981) The isolation and characterization of a cadmium- and zinc-binding protein from *Tetrahymena pyriformis*. Agric Biol Chem 45:1167–1172

Nilsson S (1964) Freshwater hyphomycetes: taxonomy, morphology and ecology. Symb Bot Ups 18:1–130

Petersen RC, Persson U (1987) Comparison of the biological effects of humic materials under acidified conditions. Sci Total Environ 62:387–398

Petersen RH (1962) Aquatic hyphomycetes from North America. I Aleuriosporae (Part 1), and key to the genera. Mycologia 54:117–151

Petersen RH (1963a) Aquatic hyphomycetes from North America. II Aleuriosporae (Part 2) and Blastosporae. Mycologia 55:18–29

Petersen RH (1963b) Aquatic hyphomycetes from North America. III Phialosporae and miscellaneous species. Mycologia 55:570–581

Pfaender FK, Alexander M (1972) Extensive microbial degradation of DDT in vitro and DDT metabolism by natural communities. J Agric Food Chem 20:842–846

Pieper HG (1978) Oekophysiologische Untersuchungen an Jugendstadien von *Gammarus fossarum*. Arch Hydrobiol (Suppl) 54:257–327

Pierce HR, Viktor MD (1978) The fate of pentachlorophenol in an aquatic ecosystem. In: Rao KR (ed) Pentachlorophenol: chemistry, pharmacology and environmental toxicology. Plenum Press, New York, pp 41–52

Pignatello J, Martinson MM, Steiert JG, Carlson RE, Crawford RL (1983) Biodegradation and photolysis of pentachlorophenol in artificial freshwater streams. Appl Environ Microbiol 46:1024–1031

Sládecková A (1963) Aquatic deuteromycetes as indicators of starch campaign pollution. Int Rev Gesamten Hydrobiol 48:35–42

Suberkropp K (1984) Effect of temperature on seasonal occurrence of aquatic hyphomycetes. Trans Br Mycol Soc 82:53–62

Suberkropp K, Michelis A, Lorch H-J, Ottow JCG (1988) Effect of sewage treatment plant effluents on the distribution of aquatic hyphomycetes in the River Erms, Schwäbische Alb, F.R.G. Aquat Bot 32:141–153

Thompson PL, Bärlocher F (1989) Effect of pH on leaf breakdown in streams and in the laboratory. J N Am Benthol Soc 8:203–210

Van der Merwe WJJ, Jooste WJ (1988) A synecological study of aquatic Hyphomycetes in the Mooi river, western Transvaal, and their significance in the decomposition of allochthonous leaf litter. S Afr J Sci 84:314–320

Van Frankenhuyzen K, Geen GH (1986) Microbe-mediated effects of low pH on availability of detrital energy to a shredder, *Clistoronia magnifica* (Trichoptera: Limnephilidae). Can J Zool 64:421–426

Weinbach EC (1956) Pentachlorophenol and mitochondrial adenosinetriphosphatase. J Biol Chem 221:609–618

10 Research on the Indian Subcontinent

K.R. Sridhar, K.R. Chandrashekar, and K.M. Kaveriappa

10.1 Introduction

Aquatic hyphomycetes were first found and described from streams, ponds and lakes in temperate regions, and research efforts into their ecology, physiology and taxonomy continue to concentrate on North American and European species (Chap. 1). The principal objective of this chapter is to summarize and discuss the research carried out on aquatic hyphomycetes in the Indian subcontinent and to suggest strategies that might make further studies more useful.

Regardless of climate and geographic location, many of the evolutionary pressures faced by aquatic hyphomycetes are likely to be the same. They include the necessity to disperse in running water, to produce enzymes active against plant polysaccharides, to survive adverse conditions such as droughts, to compete with detritus-feeding invertebrates for their food. Not surprisingly, as this chapter will show, many of the basic properties are shared by species from all geographic locations. This is most obvious when the shape of their conidia is considered. In tropical as well as in temperate species, multiradiate or sigmoid forms predominate.

There are, however, factors that clearly differ between climatic zones. Temperature regimes, overall seasonal changes (dry and wet season, summer and winter), types of substrata available for fungal colonization, timing of leaf fall, activities and types of leaf-shredding invertebrates are strongly influenced by the geographic location.

An important first step in most ecological studies is an inventory of species occurring in different habitats. Not surprisingly, this is what mycologists did first when they became interested in aquatic hyphomycetes of tropical countries. Unfortunately, many of the early reports were based on a few samples often collected more or less at random. The main conclusions that can be drawn from them is that aquatic hyphomycetes are widespread, and that there are many species that are more common in or restricted to tropical regions.

Tropical regions from which aquatic hyphomycetes have been reported include Africa (Ingold 1956, 1958, 1959, 1960; Dixon 1959; Le'John 1965;

Alasoadura 1968a,b,c; Singh 1972, 1976; Singh and Musa 1977; Bhat and Chien 1990), Australia (Shaw 1972; Shaw and Sutton 1985), Cuba (Marvanová and Marvan 1969), the Dominican Republic (Betancourt et al. 1986), the Hawaiian Islands (Anastasiou 1964; Goos 1970, 1978; Ranzoni 1979), Indonesia (Nayo 1975), Jamaica (Hudson and Sutton 1960; Hudson 1961; Crane and Dumont 1975) Malaysia (Nawawi 1985a,b, 1987; Kuthubutheen 1987; Kuthubutheen and Nawawi 1987, 1988; Nawawi and Kuthubutheen 1987, 1988a,b, 1990), Papua New Guinea and Solomon Islands (Tubaki 1965; Matsushima 1971a,b; Shaw and Sutton 1985), Puerto Rico (Padgett 1976; Betancourt and Caballero 1983; Betancourt et al. 1987; Betancourt and Justiniano 1989), Taiwan (Matsushima 1980), Thailand (Tubaki et al. 1983) and Venezuela (Nilsson 1962).

The Indian subcontinent encompasses a wide range of climatic conditions. Tropical conditions prevail; some regions have subtropical/temperate climates. Extensive forested areas with numerous permanent and temporary streams are concentrated in the Western Ghats, the Eastern Ghats and the Himalayan range (Fig. 10.1). They include alpine/subalpine, moist deciduous, evergreen and tropical rain forests. The other major biomes represented are semiarid zones in the southern and northern plains, the Thar desert in the northwest and vast estuaries on both west and east coasts.

The Western Ghats consist of a wide range of mountains, about 1600 km long. They run parallel to the west coast from the mouth of the River Tapti (21°N) to the tip of South India (8°N). They extend through the States of Maharashtra, Karnataka, Kerala and Tamil Nadu. This range is rich in tropical evergreen and moist deciduous forests having an average elevation of 1220 m and an average yearly rainfall of 250–450 cm. The mean annual temperature varies between 28 and 30 °C and average relative humidity is about 80%. Several fast flowing perennial streams and rivers (notably Kali, Cauvery, Payaswini, Nethravathi and Kunthi) originate from the midst of the Ghats and flow east into the Bay of Bengal or west into the Arabian Sea.

The Eastern Ghats are located mainly in Tamil Nadu and Andhra Pradesh. Their altitude ranges from 300–1800 m; average yearly rainfall is 120–180 cm and mean annual temperature varies between 28–32 °C. They are covered by a diverse natural vegetation. Major rivers flowing through this range include the Cauvery, Godavari and Krishna.

The Himalayan mountain range is located in Jammu/Kashmir, Himachal Pradesh, Uttar Predesh, Bihar, West Bengal and Assam. The central Himalayan region lies at an altitude of 1500–2650 m and receives annual precipitation ranging from 150–250 cm, about 75% of which falls between June and September. Average temperature is 13–15 °C. Summer months are moderately hot with a temperature of 25–30 °C during May and June. Minimum temperatures (−5 °C) are reached during January. The relative humidity fluctuates between 31% (November) and 56% (August); during rainy days it may rise up to 100%. Moderate snowfall remains on the ground

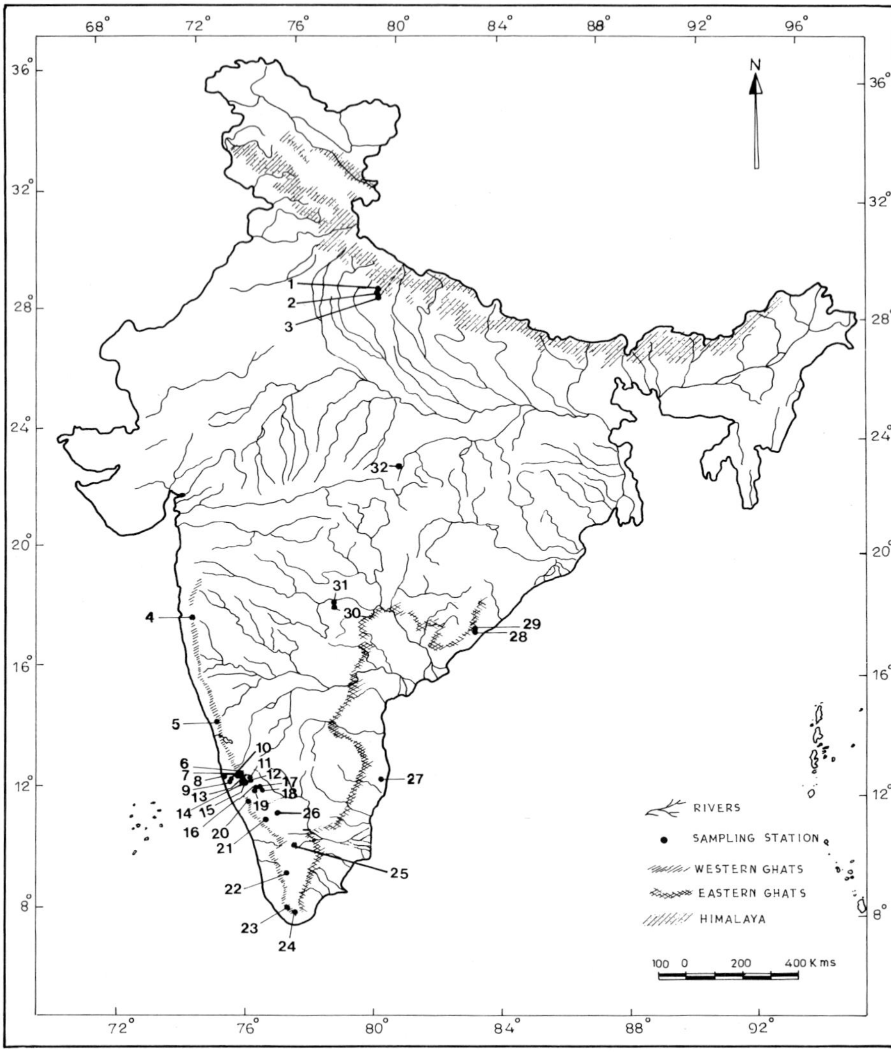

Fig. 10.1. Map of the Indian subcontinent with sampling stations (E, estuary; FL, freshwater lake; FP, freshwater pond; R, river; S, stream; TS, thermal spring) scanned for aquatic hyphomycetes. Uttar Pradesh: *1* Ram-Tal (FL); *2* Niglat (S); *3* Gufa Mahadev (S); Maharashtra: *4* Dahisar (R); Karnataka: *5* Devalamakki (S); *6* Neriya (S); *7* Uppinangadi (S); *8* Nethravathi (E); *9* Konaje (S); *10* Karaje (S); *11* Kempu Hole (R); *12* Kumaradhara (R); *13* Payaswini (R); *14* Sampaje (S); *15* Bendre Thirtha (TS); *16* Lakshmana Thirtha (S); *17* Talakaveri (S); *18* Chettalli (S); *19* Abbi Falls (S); *20* Makut (S); Kerala: *21* Kunthi (R); Tamil Nadu: *22* Thekkadi (S); *23* Mundandurai (S); *24* Kalakad (S); *25* Kodaikanal (S); *26* Mudumalai (S); *27* Kambakkam (S); Andhra Pradesh: *28* Mannanur (FP); *29* Ananthagiri (S); *30* Mancheppa (S); *31* Alisagar (S); Madhya Pradesh: *32* Jabalpur (S)

at the higher elevations during December to February. Rivers originating in the Himalaya include the Sindhu, Ganga and Brahmaputra.

10.2 Distribution

Systematic studies on aquatic hyphomycetes in India began about 40 years after Ingold's (1942) discovery of this group. Ingold and Webster (1973) were the first to report four species from a mountain stream located at Kambakkam Hills, about 80 km north of Madras (Tamil Nadu). Subsequently, these fungi were reported from other parts of India including the Himalayan region (Uttar Pradesh), the Western Ghats (Mahararashtra, Karnataka, Kerala and Tamil Nadu), the Eastern Ghats (Tamil Nadu and Andhra Pradesh) and central India (Madhya Pradesh) (see Fig. 10.1). The information on aquatic hyphomycetes found in different localities, habitats and the substrata examined for their occurrence are summarized in Table 10.1. The table also gives information on the geographical distribution of the fungi based on reports by Ingold (1975, 1976), Nilsson (1964) and Webster and Descals (1981). Only fully identified species were included.

In total, at least 32 bodies of water were surveyed for aquatic hyphomycetes. In addition to streams and rivers, a thermal spring, an estuary, freshwater ponds and a lake were examined. To date, 78 species, belonging to 45 genera have been encountered in aquatic habitats of the Indian subcontinent. Of these, 20 have been recorded only once; 24 have been found in ten or more independent surveys. Eight species have been found only in foam and two only in water samples; another six occurred predominantly in foam.

Incubation of leaf litter or examination of naturally occurring foam (Ingold 1942), foam induction (Sridhar and Kaveriappa 1984a) and water filtration (Iqbal and Webster 1973) are common methods employed to characterize aquatic hyphomycete communities of streams. Most studies reported from India are based on the examination of foam samples, and these generally yielded the highest number of species. In some studies, leaf litter of natural vegetation, plantation crops and fern was also scanned (Mer and Khulbe 1981; Sridhar 1984; Sridhar and Kaveriappa 1986a, 1988a–c, 1989e; Manoharachary and Galiah 1987). In a comparison of seeds, dried leaf litter and newspaper pieces, the seeds of *Tagetus erecta* Linn. (Mer and Khulbe 1981) and leaf midribs of banyan (*Ficus bengalensis* Linn.) (Sridhar and Kaveriappa 1989b) were found to be the best baits.

10.3 Regional Differences

Confirming observations in temperate regions of the world (Chap. 3), the aquatic hyphomycete community of Indian waters shows a marked seasonal

Table 10.1. Aquatic hyphomycetes reported from India (pp. 186–199)

Fungi	State	Habitat[a]	Source[b]	References[c]	Distribution[d]
Actinospora megalospora	Uttar Pradesh	Niglat (S,2)	F	[15]	Af,As,Eu,
Ingold	Maharashtra	? (S,?)	F	[16]	NA,SA
	Karnataka	Kempu Hole (R,11)	F	[37]	
		Payaswini (R,13)	F	[20]	
		Sampaje (S,14)	F	[34,37]	
	Kerala	Kunthi (R,21)	F	[23]	
	Andhra Pradesh	Ananthagiri (S,29)	F	[17]	
Alatospora acuminata	Uttar Pradesh	Gufa Mahadev (S,3)	F,L	[18]	Af,As,Aus,
Ingold		Niglat (S,2)	F,L	[15,19]	Eu,NA,SA
	Karnataka	Bendre Thirtha (TS,15)	F	[3]	
		Kempu Hole (R,11)	F,L,W	[2,37]	
		Konaje (S,9)	F,L,W	[22,32]	
		Lakshmana Thirtha (S,16)	L	[31]	
		Neriya (S,6)	F,L	[25,29,37]	
		Payaswini (R,13)	F,L	[20]	
		Sampaje (S,14)	F,L	[25,34,37]	
	Kerala	Kunthi (R,21)	F	[23]	
	Andhra Pradesh	Ananthagiri (S,29)	F	[4,13]	
Anguillospora crassa	Uttar Pradesh	Gufa Mahadev (S,3)	F,L	[18]	Af,As,Eu,
Ingold		Niglat (S,2)	F,L	[15, 18]	NA,SA
	Maharashtra	Dahisar (R,4)	L	[39]	
	Karnataka	Kempu Hole (R,11)	F,L,W	[1,37]	
		Kumaradhara (R,12)	F,L,W	[2]	
		Payaswini (R,13)	L	[21]	
		Sampaje (S,14)	F,L	[25,37]	
	Andhra Pradesh	Ananthagiri (S,29)	W	[11]	
		Mancheppa (S,30)/Alisagar (S,31)	F	[10]	

A. longissima (de Wild) Ingold	Uttar Pradesh	Niglat (S,2)	F	[15]	Af,As,Eu, NA,SA
	Maharashtra	Dahisar (R,4)	L	[39]	
	Karnataka	Abbi Falls (S,19)	L	[31]	
		Bendre Thirtha (TS,15)	F,L,	[3]	
		Chettalli (S,18)	L	[31]	
		Devalamakki (S,5)	F	[38]	
		Kempu Hole (R,11)	F,L,W	[1,2,37]	
		Kumaradhara (R,12)	F,L,W	[2]	
		Lakshmana Thirtha (S,16)	L	[31]	
		Nethravathi (E,8)	L	[30]	
		Payaswini (R,13)	F,L,W	[20]	
		Sampaje (S,14)	L	[27,34,37]	
		Talakaveri (R,17)	L	[31]	
	Kerala	Kunthi (R,21)	F,L,W	[23]	
	Tamil Nadu	Kalakad (S,24)	F	[38]	
		Kodaikanal (S,25)	F	[38]	
		Mudumalai (S,26)	F	[38]	
		Mundandurai (S,23)	F	[38]	
		Thekkadi (S,22)	F	[38]	
	Andhra Pradesh	Ananthagiri (S,29)	F,L,W	[4,11,17]	
A. pseudolongissima Ranzoni	Uttar Pradesh	Niglat (S,2)	F	[15]	As,NA
Articulospora angulata Tubaki	Maharashtra	Dahisar (R,4)	L	[39]	As,NA
A. inflata Ingold	Karnataka	Kempu Hole (R,11)	F	[37]	Af,As,Eu
		Payaswini (R,13)	W	[24]	
		Sampaje (S,14)	F	[37]	
A. tetracladia Ingold	Karnataka	Kempu Hole (R,11)	F,L,W	[2]	Af,As,Eu, NA,SA
		Sampaje (S,14)	F	[33]	
		Uppinangadi (S,7)	F	[38]	
	Kerala	Kunthi (R,21)	F	[38]	
	Tamil Nadu	Kalakad (S,24)	F	[38]	
		Thekkadi (S,22)	F	[38]	
	Andhra Pradesh	Ananthagiri (S,29)	W	[11]	

Table 10.1 (Continued)

Fungi	State	Habitat[a]	Source[b]	References[c]	Distribution[d]
Brachiosphaera tropicalis	Karnataka	Neriya (S,6)	F,W	[36]	Af,As,SA
Nawawi and Webster		Payaswini (R,13)	F	[21]	
		Sampaje (S,14)	F	[37]	
Campylospora chaetocladia	Uttar Pradesh	Niglat (S,2)	F,L	[15,19]	Af,As,Aus,
Ranzoni	Karnataka	Abbi Falls (S,19)	L	[31]	Eu,NA,SA
		Chettalli (S,18)	L	[31]	
		Kempu Hole (R,11)	F,L	[1,37]	
		Kumaradhara (R,12)	F,L,W	[2]	
		Lakshmana Thirtha (S,16)	L	[31]	
		Nethravathi (E,8)	L	[30]	
		Payaswini (R,13)	F,L,W	[20]	
		Sampaje (S,14)	F,L	[25,28,30,37]	
	Kerala	Kunthi (R,21)	F	[23]	
	Andhra Pradesh	Ananthagiri (S,29)	W	[11]	
		Mancheppa (S,30)/Alisagar (S,31)	L	[10]	
		Mannanur (FP,28)	L	[8]	
C. filicladia	Karnataka	Chettalli (S,18)	L	[31]	Af,As
Nawawi		Kempu Hole (R,11)	F,L,W	[2,37]	
		Makut (S,20)	F	[38]	
		Neriya (S,20)	F	[37]	
		Sampaje (S,14)	F,L	[27,28,37]	
		Uppinangadi (S,7)	F	[38]	
	Kerala	Kunthi (R,21)	F	[38]	
	Tamil Nadu	Kalakad (S,24)	F	[38]	
		Mundandurai (S,23)	F	[38]	
C. parvula	Karnataka	Neriya (S,6)	F	[37]	As,Eu
Kuzuha		Payaswini (R,13)	F	[21]	
		Sampaje (S,14)	F	[37]	
Clavariana aquatica	Uttar Pradesh	Niglat (S,2)	F	[15]	Af,As
Nawawi	Karnataka	Kempu Hole (R,11)	L,W	[2]	
		Neriya (S,6)	F,W	[34,37]	
		Sampaje (S,14)	F	[34,37]	

Clavariopsis aquatica	Uttar Pradesh	Niglat (S,2)	F,L	[15,18,19]	Af,As,Aus,
de Wildeman	Karnataka	Kempu Hole (R,11)	F	[1,37]	Eu,NA,SA
		Payaswini (S,13)	F,W	[20]	
		Sampaje (S,14)	F,L	[27,28,37]	
	Kerala	Kunthi (R,21)	F	[38]	
	Tamil Nadu	Mundandurai (S,23)	F	[38]	
	Andhra Pradesh	Ananthagiri (S,29)	L,W	[4,11,17]	
		Mancheppa (S,30)/Alisagar (S,31)	F	[10]	
		Mannanur (FP,28)	L	[8,12]	
C. azlanii	Karnataka	Kempu Hole (R,11)	F	[37]	As,SA
Nawawi		Neriya (S,6)	F	[37]	
		Sampaje (S,14)	F,L,W	[34]	
C. brachycladia	Karnataka	Sampaje (S,14)	F	[37]	As,Eu
Tubaki	Andhra Pradesh	Ananthagiri (S,29)	F,L	[4,17]	
		Mannanur (FP,28)	L	[8,12]	
C. bulbosa	Karnataka	Nethravathi (E,8)	L	[30]	As,NA
Anastasiou	Andhra Pradesh	Ananthagiri (S,29)/Mannanur (S,28)	F	[13]	
Clavatospora longibrachiata	Andhra Pradesh	?	?	[9]	As,Eu
(Ingold) Nilsson					
C. tentacula	Uttar Pradesh	Niglat (S,2)	F	[15]	Af,As,NA,
(Umphlett) Nilsson	Karnataka	Chettalli (S,18)	L	[31]	SA
		Kempu Hole (R,11)	F,L	[1,37]	
		Lakshmana Thirtha (S,16)	L	[31]	
		Neriya (S,6)	F	[37]	
		Payaswini (R,13)	F,L,W	[20]	
		Sampaje (S,14)	F,L	[27]	
	Kerala	Kunthi (R,21)	F,L,W	[23]	
	Andhra Pradesh	Ananthagiri (S,29)	W	[11]	
Condylospora spumigena	Karnataka	Kempu Hole (R,11)	F	[1]	As
Nawawi		Neriya (S,6)	F	[37]	
		Sampaje (S,14)	F,L,W	[34,37]	
		Uppinangadi (S,7)	F	[38]	
	Kerala	Kunthi (R,21)	F	[23,38]	
	Nadu	Kalakad (S,24)	F	[38]	
	Tamil	Mundandurai (S,23)	F	[38]	
		Mudumalai (S,26)	F	[38]	
		Thekkadi (S,22)	F	[38]	

Table 10.1 (Continued)

Fungi	State	Habitat[a]	Source[b]	References[c]	Distribution[d]
Culicidospora gravida Petersen	Andhra Pradesh	Ananthagiri (S,29)	W	[11]	As,Eu,NA
Dactylella microaquatica Tubaki	Andhra Pradesh	Mannanur (FP,28)	L	[12]	As,Eu,NA
D. oviparasitica Stirling and Mankau	Karnataka	Kempu Hole (R,11)	F	[37]	As,NA
		Neriya (S,6)	F	[37]	
		Sampaje (S,14)	F	[37]	
D. submersa (Ingold) Nilsson	Karnataka	Konaje (S,9)	F,L	[22]	Af,As,Eu
		Neriya (S,6)	F,L	[36]	
Dendrospora erecta Ingold	Karnataka	Konaje (S,9)	L	[27]	As,Eu,NA
		Lakshmana Thirtha (S,16)	L	[31]	
		Sampaje (S,14)	F	[37]	
D. juncicola Iqbal	Karnataka	Neriya (S,6)	F,L	[36,37]	As,Eu
Dendrosporomyces prolifer Nawawi, Webster and Davey	Karnataka	Konaje (S,9)	L	[33]	As
Erynia conica (Nowakowski) Rem. and Henn.	Karnataka	Sampaje (S,14)	F	[37]	As,Eu,NA
E. rhizospora (Thaxter) Rem. and Henn.	Karnataka	Kempu Hole (R,11)	F	[37]	As,Eu,NA
		Sampaje (S,14)	F	[37]	
Flabellocladia tetracladia (Nawawi) Nawawi	Karnataka	Neriya (S,6)	F	[37]	As
		Payaswini (R,13)	F	[21]	
		Sampaje (S,14)	F	[37]	
	Kerala	Kunthi (R,21)	F,L,W	[23]	
Flabellospora crassa Alasoadura	Uttar Pradesh	Niglat (S,2)	F	[15]	Af,As
	Karnataka	Kempu Hole (R,11)	F,L,W	[1,2,37]	
		Neriya (S,6)	F	[37]	
		Nethravathi (E,8)	L	[30]	
		Payaswini (R,13)	F,L	[20]	
		Sampaje (S,14)	F,L	[27,37]	

F. multiradiata Nawawi	Karnataka	Kempu Hole (R,11)	F	[37]	As
		Payaswini (R,13)	F	[21]	
		Sampaje (S,14)	F	[37]	
	Kerala	Kunthi (R,21)	F,L,W	[23]	
F. verticillata Alasoadura	Uttar Pradesh	Niglat (S,2)	L	[19]	Af,As
	Maharashtra	? (S,?)	F	[16]	
	Karnataka	Nethravathi (E,8)	L	[30]	
		Bendre Thirtha (TS,15)	L	[3]	
		Devalamakki (S,5)	F	[38]	
		Kempu Hole (R,11)	F	[37]	
		Konaje (S,9)	F	[22]	
		Neriya (S,6)	F,L	[25,29]	
		Payaswini (R,13)	F,L	[20]	
		Sampaje (S,14)	F,L	[25,34,37]	
	Kerala	Kunthi (R,21)	F,L,W	[23]	
	Tamil Nadu	Mudumalai (S,26)	F	[38]	
		Mundandurai (S,23)	F	[38]	
	Andhra Pradesh	Ananthagiri (S,29)	F,L	[4,17]	
Flagellospora curvula Ingold	Karnataka	Abbi Falls (S,19)	L	[31]	Af,As,Eu, NA,SA
		Bendre Thirtha (TS,15)	F,L,W	[3]	
		Chettalli (S,18)	L	[31]	
		Kempu Hole (R,11)	F,L,W	[1,37]	
		Konaje (S,9)	F,L,W	[22]	
		Kumaradhara (R,12)	F,L,W	[2]	
		Lakshmana Thirtha (S,16)	L	[31]	
		Neriya (S,6)	F,L	[33,37]	
		Nethravathi (E,8)	L	[30]	
		Payaswini (R,13)	F,L	[21]	
		Sampaje (S,14)	F,L	[27,30,34,37]	
		Talakaveri (S,17)	L	[31]	
	Kerala	Kunthi (R,21)	F,L,W	[23]	

Table 10.1 (Continued)

Fungi	State	Habitat[a]	Source[b]	References[c]	Distribution[d]
F. penicillioides Ingold	Uttar Pradesh	Gufa Mahadev (S,3)	L	[18]	Af,As,Aus, Eu,NA,SA
		Niglat (S,2)	F,L	[15,18]	
	Karnataka	Abbi Falls (S,19)	L	[31]	
		Bendre Thirtha (TS,15)	F,L,W	[3]	
		Kempu Hole (R,11)	F,L,W	[1,37]	
		Konaje (S,9)	F,L,W	[22,32]	
		Kumaradhara (R,12)	F,L,W	[2]	
		Lakshmana Thirtha (S,16)	L	[31]	
		Neriya (S,6)	F,L	[33,37]	
		Nethravathi (E,8)	L	[30]	
		Payaswini (R,13)	F,L,W	[20]	
		Sampaje (S,14)	F,L	[28,29,30,37]	
		Talakaveri (S,17)	L	[31]	
	Kerala	Kunthi (R,21)	W	[23]	
	Andhra Pradesh	Ananthagiri (S,29)	L	[4,17]	
		Mannanur (FP,28)	L	[8,12]	
F. prolifera Petersen	Madhya Pradesh	Jabalpur (S,32)	F	[5]	As,NA
Geniculospora inflata (Ingold) Nils. ex Marv. and Nilsson	Andhra Pradesh	?	?	[9]	As,Eu
Gyoerffyella tricapillata (Ingold) Marvanová	Maharashtra	? (FP,?)	S	[16]	As,Eu
	Andhra Pradesh	?	?	[9]	
Heliscella stellata (Ingold and Cox) Marvanová	Karnataka	Lakshmana Thirtha (S,16)	L	[31]	As,Eu,NA
		Payaswini (R,13)	W	[24]	
		Sampaje (S,14)	F	[37]	
Heliscus lugdunensis Sacc. and Théry	Uttar Pradesh	Gufa Mahadev (S,3)	L	[18]	As,Aus, Eu,NA
	Andhra Pradesh	?	?	[9]	
H. submersus Hudson	Andhra Pradesh	?	?	[9]	As,SA

Ingoldiella hamata Shaw	Karnataka	Kempu Hole (R,11)	F,L,W	[1,37]	Af,As,Aus
		Konaje (S,9)	F,L,W	[22,32]	
		Kumaradhara (R,12)	F,L,W	[2]	
		Neriya (S,6)	F,L	[25,29,37]	
		Payaswini (R,13)	F	[20]	
		Sampaje (S,14)	F,L	[25,27,30,37]	
	Tamil Nadu	Kambakkam (S,27)	F	[7]	
	Andhra Pradesh	? (S,?)	F	[6]	
Isthmotricladia gombakiensis Nawawi	Karnataka	Kempu Hole (R,11)	F	[37]	Af,As
		Neriya (S,6)	F,L	[27,29,37]	
		Nethravathi (E,8)	L	[30]	
		Payaswini (R,13)	F	[21]	
		Sampaje (S,14)	F	[37]	
	Andhra Pradesh	Ananthagiri (S,29)	L	[17]	
I. laeensis Matsushima	Karnataka	Chettalli (S,18)	L	[31]	Af,As
		Kempu Hole (R,11)	F	[37]	
		Kumaradhara (R,12)	F,L,W	[2]	
		Lakshmana Thirtha (S,16)	L	[31]	
		Neriya (S,6)	F,L	[29,37]	
		Payaswini (R,13)	L	[20]	
		Sampaje (S,14)	F,L	[27,28,37]	
Jaculispora submersa Hudson and Ingold	Andhra Pradesh	Mannanur (FP,28)	L	[12]	Af,As,Eu, NA,SA
Laridospora appendiculata (Anastasiou) Nawawi	Maharashtra	Dahisar (R,4)	L	[39]	Af,As,Eu
	Karnataka	Kempu Hole (R,11)	F,L,W	[1,2,37]	
		Konaje (S,9)	F,L	[33]	
		Lakshmana Thirtha (S,16)	L 27	[31]	
		Neriya (S,6)	F,L	[27,29,37]	
		Uppinangadi (S,7)	F	[38]	
	Kerala	Kunthi (R,21)	F	[38]	
	Tamil Nadu	Kalakad (S,24)	F	[38]	
		Mudumalai (S,26)	F	[38]	
		Mundandurai (S,23)	F	[38]	
		Thekkadi (S,22)	F	[38]	

Table 10.1 (Continued)

Fungi	State	Habitat[a]	Source[b]	References[c]	Distribution[d]
Lateriramulosa uni-inflata Matsushima	Maharashtra	? (S,?)	F	[16]	Af,As
	Karnataka	Kempu Hole (R,11)	F,L,W	[2,37]	
		Konaje (S,9)	F,L	[22]	
		Neriya (S,6)	F,L	[25,29,37]	
		Payaswini (R,13)	L	[20]	
		Sampaje (S,14)	F,L	[25,34,37]	
		Uppinangadi (S,7)	F	[38]	
	Kerala	Kunthi (R,21)	F	[38]	
	Tamil Nadu	Mundandurai (S,23)	F	[38]	
	Andhra Pradesh	Ananthagiri (S,29)	F	[4,17]	
Lemonniera aquatica de Wildeman	Uttar Pradesh	Niglat (S,2)	F,L	[14,15]	Af,As,Aus, Eu,NA,SA
		Ram-Tal (FL,1)	L	[14]	
	Maharashtra	Dahisar (R,4)	L	[39]	
	Andhra Pradesh	?	?	[9]	
	Madhya Pradesh	Jabalpur (S,32)	?	[5]	
L. centrosphaera Marvanová	Andhra Pradesh	Ananthagiri (S,29)	W	[11]	Af,As,Eu
Lunulospora curvula Ingold	Uttar Pradesh	Niglat (S,2)	F,L	[14,15,19]	Af,As,Aus, Eu,NA,SA
		Ram-Tal (FL,1)	L	[14]	
	Karnataka	Abbi Falls (S,19)	L	[31]	
		Bendre Thirtha (TS,15)	F,L,W	[3]	
		Chettalli (S,18)	L	[31]	
		Kempu Hole (R,11)	F,L,W	[1,37]	
		Konaje (S,9)	F,L,W	[22,32]	
		Kumaradhara (R,12)	F,L,W	[2]	
		Lakshmana Thirtha (S,16)	L	[31]	
		Neriya (S,6)	F,L	[25,27,29,37]	
		Nethravathi (E,8)	L	[30]	
		Payaswini (R,13)	F,L,W	[20]	
		Sampaje (S,14)	F,L	[25,30,34,37]	
		Talakaveri (S,14)	L	[31]	

	Kerala	Kunthi (R,21)	F,L,W	[23]	
	Tamil Nadu	Kambakkam (S,27)	L	[7]	
	Andhra Pradesh	Ananthagiri (S,29)	F,L,W	[4,11,17]	
		Mancheppa (S,30)/Alisagar (S,31)	F	[10]	
		Mannanur (FP,28)	L	[8,12]	
L. cymbiformis Miura	Uttar Pradesh	Niglat (S,2)	L	[19]	Af,As,Aus
	Karnataka	Bendre Thirtha (TS,15)	L,W	[3]	
		Kempu Hole (R,12)	F,L,W	[1,37]	
		Kumaradhara (R,12)	F,L,W	[2]	
		Neriya (S,6)	F	[37]	
		Nethravathi (E,8)	L	[30]	
		Payaswini (R,13)	F,L,W	[20]	
		Sampaje (S,14)	F,L	[27,28,34,37]	
		Talakaveri (S,17)	L	[31]	
	Kerala	Kunthi (R,21)	F,L,W	[23]	
	Andhra Pradesh	Ananthagiri (S,29)	W	[11]	
Margaritispora aquatica Ingold	Andhra Pradesh	Ananthagiri (S,29)	W	[11]	As,Eu,NA
Nawawia filiformis (Nawawi) Marvanová	Karnataka	Kempu Hole (R,11)	F	[2,37]	Af,As,Eu
		Sampaje (S,14)	F	[37]	
		Uppinangadi (S,7)	L	[38]	
	Kerala	Kunthi (R,21)	F	[38]	
	Tamil Nadu	Kalakad (24)	F	[38]	
Phalangispora constricta Nawawi and Webster	Karnataka	Bendre Thirtha (TS,15)	F,L	[3]	Af,As
		Kempu Hole (R,11)	F	[37]	
		Konaje (S,9)	F,L,W	[22,32,35]	
		Kumaradhara (R,12)	F,L,W	[2]	
		Neriya (S,6)	F,L	[29,37]	
		Payaswini (R,13)	F,L,W	[20,21]	
		Sampaje (S,14)	F,L	[28,37]	
	Kerala	Kunthi (R,21)	F,L,W	[23]	
	Andhra Pradesh	Ananthagiri (S,29)	L	[4,17]	
P. nawawii Kuthubutheen	Karnataka	Sampaje (S,14)	F	[37]	As

Table 10.1 (Continued)

Fungi	State	Habitat[a]	Source[b]	References[c]	Distribution[d]
Pyramidospora casuarinae	Karnataka	Konaje (S,9)	L	[25,32]	Af,As,SA
Nilsson		Sampaje (S,14)	F,L	[37]	
P. constricta	Karnataka	Konaje (S,9)	F,L,W	[22,27]	Af,As
Singh		Neriya (S,6)	F,L	[27,37]	
		Sampaje (S,14)	F	[37]	
Speiropsis hyalospora	Maharashtra	? (S,?)	F	[16]	As
Subramanian and Lodha	Karnataka	Kempu Hole (R,11)	F	[37]	
		Neriya (S,6)	F,L	[25,29,37]	
		Payaswini (R,13)	L	[24]	
		Sampaje (S,14)	F,L	[25,34,37]	
	Kerala	Kunthi (R,21)	F	[23]	
S. irregularis	Karnataka	Kempu Hole (R,11)	F	[37]	As,NA
Petersen		Sampaje (S,14)	F	[37]	
	Andhra Pradesh	Ananthagiri (S,29)	F,L	[4,17]	
S. pedatospora	Maharashtra	? (S,?)	F	[16]	Af,As
Tubaki	Karnataka	Bendre Thirtha (TS,15)	L	[3]	
		Kempu Hole (R,11)	F,L,W	[2,37]	
		Neriya (S,6)	F,L	[25,29,37]	
		Payaswini (R,13)	F	[21]	
		Sampaje (S,14)	F,L	[25,28,30,37]	
	Andhra Pradesh	?	?	[9]	
Tetrachaetum elegans	Uttar Pradesh	Niglat (S,2)	F,L	[15,18,19]	Af,As,Aus,
Ingold	Andhra Pradesh	Ananthagiri (S,29)	F	[4,17]	Eu,NA
Tetracladium marchalianum	Uttar Pradesh	Niglat (S,2)	F,L	[14,15]	Af,As,Aus,
de Wildeman		Ram-Tal (FL,1)	L	[14]	Eu,NA,SA
	Maharashtra	? (S,?)	F	[16]	
	Karnataka	Chettalli (S,18)	L	[31]	
		Kempu Hole (R,11)	F,L	[1,2,37]	
		Kumaradhara (R,12)	F,L,W	[2]	
		Lakshmana Thirtha (S,16)	L	[31]	
		Neriya (S,6)	F	[37]	
		Sampaje (S,14)	F	[33,37]	
	Andhra Pradesh	?	?	[9]	

T. setigerum (Grove) Ingold	Uttar Pradesh	Niglat (S,2)	F,L	[14,15]	Af,As,Eu, NA
		Ram-Tal (FL,1)	L	[14]	
	Maharashtra	Dahisar (R,4)	L	[39]	
		? (S,?)	F	[16]	
	Karnataka	Devalamakki (S,5)	F	[38]	
		Kempu Hole (R,11)	F	[37]	
		Sampaje (S,14)	F	[37]	
		Uppinangadi (S,7)	F	[38]	
	Kerala	Kunthi (R,21)	F	[38]	
	Tamil Nadu	Kalakad (S,24)	F	[38]	
		Mundandurai (S,23)	F	[38]	
		Thekkadi (S,22)	F	[38]	
	Andhra Pradesh	?	?	[9]	
Tricellula aquatica Webster	Andhra Pradesh	?	?	[9]	As,Eu,NA
Tricladiomyces malaysianum (Nawawi) Nawawi	Karnataka	Karaje (S,10)	L	[36]	As
Tricladiospora brunnea (Nawawi) Nawawi and Kuthubutheen	Karnataka	Bendre Thirtha (TS,15)	F	[3]	As
		Kempu Hole (R,11)	F	[37]	
		Konaje (S,9)	F,L	[22]	
		Neriya (S,6)	F,L	[29,37]	
		Sampaje (S,14)	F,L	[33,37]	
Tricladium angulatum Ingold	Uttar Pradesh	Niglat (S,2)	F	[15]	Af,As,Eu, NA
	Maharashtra	? (S,?)	F	[16]	
	Karnataka	Neriya (S,6)	F,L,W	[36]	
T. anomalum Ingold	Uttar Pradesh	Niglat (S,2)	F	[15]	Af,As
T. fuscum Nawawi	Karnataka	Karaje (S,10)	W	[36]	As
T. splendens Ingold	Uttar Pradesh	Niglat (S,2)	L	[19]	Af,As,Eu, NA
	Maharashtra	? (S,?)	?	[16]	
	Andhra Pradesh	Ananthagiri (S,29)	W	[11]	
Tripospermum camelopardus Ingold, Dann and McDougall	Tamil Nadu	Mundandurai (S,23)	F	[38]	As,Eu

Table 10.1 (Continued)

Fungi	State	Habitat[a]	Source[b]	References[c]	Distribution[d]
T. myrti (Lind.) Hughes	Karnataka	Abbi Falls (S,19)	L	[31]	As,Eu
		Chettalli (S,27)	L	[31]	
		Payaswini (R,13)	F,L	[20]	
		Sampaje (S,14)	F,L	[25,28]	
	Kerala	Kunthi (R,21)	F	[23]	
	Tamil Nadu	Kalakad (S,24)	F	[38]	
Triscelophorus acuminatus Nawawi	Uttar Pradesh	Niglat (S,2)	F	[15]	Af,As
	Karnataka	Bendre Thirta (TS,15)	F,L,W	[3]	
		Kempu Hole (R,11)	F,L,W	[1,37]	
		Konaje (S,9)	F,L,W	[22,32]	
		Kumaradhara (R,12)	F,L,W	[2]	
		Lakshmana Thirtha (S,16)	L	[31]	
		Neriya (S,6)	F,L	[27,29,37]	
		Nethravathi (E,8)	L	[30]	
		Payaswini (R,13)	F,L,W	[20]	
		Sampaje (S,14)	F,L	[27,28,30,37]	
		Talakaveri (S,17)	L	[31]	
	Kerala	Kunthi (R,21)	F,L,W	[23]	
T. konajensis Sridhar and Kaveriappa	Karnataka	Kempu Hole (R,11)	F	[37]	As
		Konaje (S,9)	F,L,W	[26,32,34]	
		Neriya (S,6)	F,L	[35,37]	
		Sampaje (S,14)	F,L,W	[34]	
T. monosporus Ingold	Uttar Pradesh	Niglat (S,2)	F,L	[14,15,19]	Af,As,Aus, Eu,NA,SA
		Ram-Tal (FL,1)	L	[14]	
	Maharashtra	? (S,?)	F	[16]	
	Karnataka	Abbi Falls (S,19)	L	[31]	
		Bendre Thirtha (TS,15)	F,L,W	[3]	
		Chettalli (S,18)	L	[31]	
		Kempu Hole (R,11)	F,L,W	[1,37]	
		Konaje (S,9)	L	[32]	
		Kumaradhara (R,12)	F,L,W	[2]	
		Lakshmana Thirtha (S,16)	L	[31]	

		Neriya (S,6)	F,L	[25,29,37]	
		Nethravathi (E,8)	L	[30]	
		Payaswini (R,13)	F,L,W	[20]	
		Sampaje (S,14)	F,L	[25,28,30,37]	
		Talakaveri (S,17)	L	[31]	
	Kerala	Kunthi (R,21)	F,L,W	[23]	
	Tamil Nadu	Kambakkam (S,27)	F	[7]	
	Andhra Pradesh	Ananthagiri (S,29)	F,L	[4,13,17]	
		Mancheppa (S,30)/Alisagar (S,31)	F,W	[10,11]	
		Mannanur (FP,28)	L	[8,12]	
Trisulcosporium acerinum Hudson and Sutton	Andhra Pradesh	Ananthagiri (S,29)	W	[11]	Af,As,Eu
Tumularia aquatica (Ingold) Descals and Marvanová	Karnataka	Kempu Hole (R,11)	F,W	[1,2,37]	Af,As,Eu, NA,SA
		Neriya (S,6)	F	[37]	
		Payaswini (R,13)	W	[23]	
		Sampaje (S,14)	F	[37]	
	Andhra Pradesh	Mancheppa (S,30)/Alisagar (S,31)	F	[10]	
Varicosporium elodeae Kegel	Karnataka	Konaje (S,9)	F	[22]	Af,As,Eu, NA
		Lakshmana Thirtha (S,16)	L	[31]	
		Sampaje (S,14)	F	[37]	
	Andhra Pradesh	Ananthagiri (S,29)	W	[11]	

[a] *E*, Estuary; *FL*, Freshwater Lake; *FP*, Freshwater Pond; *R*, River; *S*, Stream; *TS*, Thermal Spring (number in parenthesis refers to the location in the map).
[b] *F*, Foam; *L*, Leaf litter; *S*, Scum; *W*, Water.
[c] [1] Chandrashekar et al. (1986); [2] Chandrashekar et al. (1990); [3] Chandrashekar et al. (1991); [4] Galiah and Manoharachary (1987); [5] Hasija and Shanware (1986); [6] Ingold (1973); [7] Ingold and Webster (1973); [8] Madhusudan Rao and Manoharachary (1984); [9] Manoharachary (1989); [10] Manoharachary and Bhairavanath (1985); [11] Manoharachary and Galiah (1987); [12] Manoharachary and Madhusudan Rao (1983); [13] Manoharachary and Murthy (1981); [14] Mer and Khulbe (1981); [15] Mer and Sati (1989); [16] Patil and Kapadnis (1980); [17] Sarma and Manoharachary (1989); [18] Sati and Tiwari (1990); [19] Sati et al. (1989); [20] Sridhar and Kaveriappa (1982); [21] Sridhar and Kaveriappa (1984a); [22] Sridhar and Kaveriappa (1984b); [23] Sridhar and Kaveriappa (1985a); [24] Sridhar and Kaveriappa (1985b); [25] Sridhar and Kaveriappa (1986a); [26] Sridhar and Kaveriappa (1987a); [27] Sridhar and Kaveriappa (1988a); [28] Sridhar and Kaveriappa (1988b); [29] Sridhar and Kaveriappa (1988c); [30] Sridhar and Kaveriappa (1988d); [31] Sridhar and Kaveriappa (1989a); [32] Sridhar and Kaveriappa (1989b); [33] Sridhar and Kaveriappa (1989c); [34] Sridhar and Kaveriappa (1989d); [35] Sridhar and Kaveriappa (1989e); [36] Sridhar and Kaveriappa (1990); [37] Sridhar and Kaveriappa (1992); [38] Subramanian and Bhat (1981); [39] Thakur (1977).
[d] Af, Africa; As, Asia; Aus, Australia; Eu, Europe; NA, North America; SA, South America.

periodicity. The main factors that appear to be correlated with these fluctuations are water temperature, timing of litter fall and precipitation, changes between wet and dry seasons and the types of substrata. To analyze the factors influencing aquatic hyphomycete communities, the studies in three different geographical areas viz., the Central Himalayas, the Western Ghats and the Eastern Ghats were considered. In the Central Himalayan region (Mer and Sati 1989), the highest number of fungal species was recorded during the monsoon season (September–November) and during spring (March–April). During these periods, the water temperature fluctuates between 14–19 °C. This might have a major influence on the fungi, especially on their sporulation. At other times of the year, water temperature is lower (8–13 °C from December to February) or higher (21–25 °C from May to July). Leaf fall is highest during winter (December–February). In addition, heavy rains during the monsoon season will carry considerable amounts of leaves into the streams and rivers.

The situation in the Western and Eastern Ghats seems to be similar as far as maxima of conidial numbers and types are concerned. They are generally highest during the later part of the monsoon season (August–November). At that time, the water temperature will be 17–22 °C. Leaves are shed from March through June; rainfall begins in June/July and continues through November (Manoharachary and Galiah 1987; Sridhar and Kaveriappa 1989a,d).

In coastal streams (west coast), the number of species and conidia transported in the water column is greater from July to December than from January to June (Sridhar and Kaveriappa 1984b). The same pattern was observed with leaf baiting experiments (Sridhar and Kaveriappa 1989b). The number of species per sample varied between 6–20 in coastal streams compared to 24–36 in Western Ghat streams (Sridhar and Kaveriappa 1984b, 1989b; Chandrashekar et al. 1990). The lower numbers in coastal streams might be due to higher water temperatures (23–30 °C), a longer dry season (mid December to end of May; February to May in Western Ghats), and comparatively sparse riparian vegetation (Sridhar and Kaveriappa 1984b, 1989b). Rainfall, litter input, water turbulence and dissolved oxygen are always higher in Western Ghat streams than in coastal streams, whereas the water temperature is much lower in Western Ghat streams (Sridhar and Kaveriappa 1984b, 1989a,b,d). A dense canopy of forests, plantation crops and higher altitude are mainly responsible for keeping the forest floor as well as the streams comparatively cool.

No data are available to compare the coastal or Western Ghat hyphomycete communities with those of the plains in semiarid zones.

10.4 Substrata

Generally, there are no clearcut substratum preferences in aquatic hyphomycetes. Nevertheless, colonization patterns appear to be influenced by leaf substances that selectively inhibit or stimulate the growth of certain species, by differences in the enzymatic equipment of the fungi, and by competition among them.

Fungal colonization patterns on submerged leaves of eight plantation crops were followed over a period of one year in the Sampaje stream in the Western Ghats (unpubl. obs.). Overall, 27 fungal species were recorded. Coffee (*Coffea arabica* Linn.) and rubber (*Hevea brasiliensis* M.) leaves were colonized by 22 and 20 species respectively; whereas only four species colonized jack (*Artocarpus heterophyllus* Lam.) leaves. An unknown synnematous fungus was dominant, followed by *Triscelophorus konajensis*, *T. acuminatus*, *T. monosporus*, *Flagellospora penicillioides* and *Lunulospora curvula* (Table 10.2).

Table 10.2. Colonization of leaves of plantation crops by aquatic hyphomycetes

Fungus	Leaf species[a]								Frequency of occurrence[b]
	1	2	3	4	5	6	7	8	
Unknown (synnematous fungus)	+	+	+	+	+	+	+	+	248
Triscelophorus konajensis Sridhar and Kaveriappa	+	+	+		+	+	+	+	236
T. acuminatus Nawawi	+	+	+		+	+	+	+	215
T. monosporus Ingold	+	+	+	+	+	+	+		203
Flagellospora penicillioides Ingold	+	+	+		+	+	+		193
Lunulospora curvula Ingold	+	+	+	+	+		+	+	162
Phalangispora constricta Nawawi and Webster	+		+	+					111
Flabellospora verticillata Alasoadura	+	+	+	+					92
Campylospora chaetocladia Ranzoni	+	+				+			87
Lunulospora cymbiformis Miura	+	+				+			78
Speiropsis pedatospora Tubaki	+	+	+	+		+			61
Ingoldiella hamata Shaw	+	+	+		+				53
Anguillospora longissima (de Wild.) Ingold	+	+							37
Lateriramulosa uni-inflata Matsushima	+	+							31
Speiropsis hyalospora Subramanian and Lodha	+	+	+	+					28
Alatospora acuminata Ingold	+	+							27
Flabellospora multiradiata Nawawi				+			+		23

Table 10.2 (Continued)

Fungus	Leaf species[a]								Frequency of occurrence[b]
	1	2	3	4	5	6	7	8	
Anguillospora crassa Ingold	+	+							19
Clavatospora tentacula (Umphlett) Nilsson	+	+							12
Flagellospora curvula Ingold	+	+							11
Isthmotricladia laeensis Matsushima	+								9
Tripospermum myrti (Lind.) Hughes		+							7
Tricladiospora brunnea (Nawawi) Nawawi and Kuthubutheen					+				5
Campylospora filicladia Nawawi	+								5
Clavariopsis aquatica De Wildeman		+							3
Isthmotricladia gombakiensis Nawawi				+					3
Flabellospora crassa Alasoadura	+								1
Total number of species	22	20	11	9	8	8	7	4	

[a] 1, Coffee; 2, rubber; 3, mango; 4, teak; 5, cashew; 6, cinnamon; 7, coconut (part of leaflet); 8, Jack. +, Fungus was present on at least 1 of 40 leaves examined.
[b] Out of a maximum of 320 (40 leaves × 8 types).

In a study in the Neriya stream at the foot-hills of the Western Ghats submerged leaf litter of natural vegetation and plantation crops was examined. Banyan leaves were colonized by 12 species, banana (*Musa paradisiaca* Linn.) and jack leaves by only four and three species (Sridhar and Kaveriappa 1988c). Other studies confirmed that banyan leaves are among the best substrata for recovering a maximum number of species (Sridhar and Kaveriappa 1989b).

There are very few reports on aquatic hyphomycetes on exotic plant litters viz., *Acacia*, *Casuarina*, *Eucalyptus*, *Pinus* and *Rubus* (Nilsson 1962; Cowling and Waid 1963; Swart 1986; Thomas et al. 1989). The study of submerged litter from *Eucalyptus tereticornis* Smith in an Eastern Ghat stream (Manoharachary and Galiah 1987) and *Pinus roxburghii* Sarg. in a Himalayan stream (Sati et al. 1989) yielded two and nine fungal species, respectively. Eleven species were found on submerged litter of *Gleichenia pectinata* Presl., one of the common ferns in the Western Ghat region (Sridhar and Kaveriappa 1988c).

Aquatic hyphomycetes were also cultured on leaf petioles (Sridhar and Kaveriappa 1983) and plant latex (Sridhar and Kaveriappa 1987b). These substrata stimulate sporulation in pure cultures and may also be helpful in inducing the perfect state.

10.5 Survival Outside Streams

The ability to survive desiccation may be essential in ensuring the continued existence of aquatic hyphomycetes in a given region (Chap. 2). It seems likely that there are vast differences in the duration and timing of dry seasons between and within broad geographic regions. Due to elevated temperatures, the water content of litter on the ground or on dried stream beds may well be lower in tropical areas. Other factors, such as the presence of a canopy, or of a dense litter layer on the forest floor, may provide partial protection against the sun's heat.

There are few studies on the survival of aquatic hyphomycetes in tropical terrestrial situations. A total of 27 species were recorded on ten types of dried leaf litter on the banks of Western Ghat streams collected in May (Sridhar and Kaveriappa 1987c). Generally, this month corresponds to the peak of summer in the Western Ghats; the monsoon starts abruptly in early June. The atmospheric temperature in May varies between 28 and 34 °C. An unknown synnematous fungus (*Anguillospora* sp.), *Lunulospora curvula, Triscelophorus acuminatus* and *T. konajensis* were most frequently observed. On dried leaves of rubber, coffee, teak (*Tectona grandis* Linn.) and *Gleichenia* 16, 11, 9 and 9 species were encountered; from dried leaves of banana, coconut (*Cocos nucifera* Linn.) and cinnamon (*Cinnamomum zeylanicum* Blume) only four, four and two species were recovered. Coffee and rubber leaves submerged in a Western Ghat stream were colonized by 21 species of aquatic hyphomycetes; these were able to survive under dry laboratory conditions for periods ranging from 30 to over 360 days (Sridhar and Kaveriappa 1988e).

Thakur (1977) examined submerged leaves of *Pongamia pinnata* Merr. and *Polygonum glabrum* Willd. collected from the River Dahisar of Maharashtra for the presence of aquatic hyphomycetes. The leaves were then dried and kept in the laboratory for up to six months (26–30 °C). *Laridospora appendiculata, Anguillospora crassa* and *A. longissima* survived over 100 days on half-skeletonized leaves and over 40 days on skeletonized leaves. *Articulospora angulata, Lemonniera aquatica* and *Tetracladium setigerum* survived for shorter periods of time.

Ingoldiella hamata, Lunulospora curvula and *Triscelophorus konajensis* inoculated on sterilized leaf disks of banyan, cashew (*Anacardium occidentale* Linn.), coffee and rubber survived under dry laboratory conditions for a period of 30 to 90 days and on garden soil for 30 to 60 days (Sridhar and Kaveriappa, unpubl. obs.). *Articulospora tetracladia* was recorded in a terrestrial habitat as dark powdery colonies on damp bamboo wood in Poona, Maharashtra (Patil and Rao 1972).

Many leaves, together with the fungi they carry, will be swept downstream and eventually become exposed to brackish water (Chap. 2). When naturally colonized coffee and rubber leaves collected from a Western Ghat stream were incubated at different concentrations of sterile seawater, 14 species of aquatic hyphomycetes were able to sporulate at the range of 3.4–17 ppt salinity. Litter collected from estuarine waters was incubated in filtered brackish water as well as freshwater. More fungi sporulated in freshwater. Sporulation in brackish water increased during the monsoon season, when salinity and temperature decreased (July–September). There is evidence that growth and degradative activities of aquatic hyphomycetes are less vulnerable to increased salinity than reproduction (Sridhar and Kaveriappa 1988d).

10.6 Other Studies

The number of aquatic hyphomycetes in a sulfur spring in the Western Ghat region (Chandrashekar et al. 1991) was greatly reduced. They were entirely absent in the spring proper (36.5 °C; 3.1 mg sulfide l^{-1}), where only one terrestrial species was found. Approximately 150 m downstream (35.5 °C; 1.2 mg sulfide), 14 species were found. This number increased to 20 at the junction of the spring with a nearby rivulet (26.5 °C, 0.1 mg sulfide). The scarcity of aquatic hyphomycetes in the sulfur spring could be due to high temperature, high sulfide content or low oxygen content, or to a combination of these factors. For comparison, when coffee leaves collected from a Western Ghat stream were incubated, *Flabellospora verticillata, F. penicillioides* and *Helicosporium* sp. could sporulate even at 34 °C.

Flagellospora penicillioides, *Helicosporium sp.*, *Ingoldiella hamata*, *Lunulospora curvula*, *Triscelophorus acuminatus*, *T. konajensis* and *Wiesneriomyces laurinus* (Tassi) Kirk isolated from Western Ghat streams were scanned for cellulase, amylase, triacyl glycerol lipase, pyrocatechol oxidase and proteinase activity (Chandrashekar and Kaveriappa 1988). *Ingoldiella hamata* and *Phalangispora constricta* exhibited only pyrocatechol oxidase activity, whereas *Triscelophorus acuminatus* showed only amylase activity. *Helicosporium* sp. and *I. hamata* did not show triacyl glycerol lipase activity. Maximum cellulase and amylase activity were observed in *L. curvula*, whereas tryacyl glycerol lipase activity was highest in *P. constricta*.

Carboxymethyl cellulose and ammonium sulfate were the best carbon and nitrogen sources to induce cellulase production by 12-day old *Lunulospora curvula* and *Flagellospora penicillioides* (pH optimum 5.2, temperature optimum 28 °C; Chandrashekar and Kaveriappa 1991).

Amylase production by 12-day-old cultures of *Lunulospora curvula* and *Phalangispora constricta* was highest when starch and ammonium sulfate were supplied (pH optimum 5.2, temperature optimum 28 °C; Chandrashekar 1987).

Rubber leaves colonized by seven species of aquatic hyphomycetes were offered to the fish *Sarotherodan mossambica* (Chandrashekar et al. 1989). Colonized leaves were preferred over sterile leaves. Leaves colonized by *Lunulospora curvula* and *Flagellospora penicillioides* were the most palatable. Juvenile fish and prawns preferred leaf disks (*Aglaia elaegnoidea* (Juss.) Benth.) colonized by aquatic hyphomycetes over uncolonized disks (Madhusudan Rao and Manoharachary 1982).

The effect of two herbicides, namely, Paraquat (Gramaxone) and 2,4-dichlorophenoxy butyricacid (2,4-DB Fernoxone) and two fungicides viz., Dithane M-45 (Mancozeb) and Captofol (Foltop) on the growth of *Flagellospora penicillioides*, *Lunulospora curvula* and *Phalangispora constricta* was studied by Chandrashekar and Kaveriappa (1989). Fungal growth was not inhibited up to a concentration of $5\,mg\,l^{-1}$. There was slightly enhanced growth at concentrations of $1–25\,mg\,l^{-1}$ of Paraquat and 2,4-DB Fernoxone.

Sridhar and Kaveriappa (1986b) studied the effect of the two fungicides, bavistin and Bordeaux mixture, and the insecticide benzene hexachloride (BHC) on the sporulation of aquatic hyphomycetes. *Campylospora chaetocladia*, *Campylospra filicladia*, *Flabellospora verticillata Flagellospora curvula*, *F. penicillioides*, *Lunulospora curvula*, *L. cymbiformis*, *Triscelophorus acuminatus*, *Triscelophorus konajensis* and *Wiesneriomyces laurinus* were very sensitive to these chemicals at concentrations between 5 and $10\,mg\,l^{-1}$.

10.7 Conclusions

Our knowledge of aquatic hyphomycetes has greatly benefited from the introduction of innovative and sophisticated techniques such as mapping of individual leaves (Shearer and Lane 1983), biochemical studies (Chaps. 4, 6, 8), and electron microscopy (Chap. 7). By contrast, a decade of research in India has been largely confined to exploratory surveys, while ecology, taxonomy and physiology have been neglected. Close to 300 species have been described from all over the world (L. Marvanová 1991, pers. comm.); approximately one fourth of them (78 species) have been reported from India, 80% (62 species) of which occur in Western Ghat streams and rivers.

Several species that are most abundant in temperate regions nevertheless occur in tropical climates. These include *Culicidospora gravida*, *Dendrospora erecta*, *Flagellospora prolifera*, *Heliscella stellata*, *Heliscus lugdunensis*, *Margaritispora aquatica*, *Tricellula aquatica*, and *Varicosporium elodeae*. However, most of these have been recorded only sporadically and mostly, sometimes exclusively, in foam or water samples of mountain streams. This suggests that their site of reproduction must have been upstream. Even though the Western Ghats, Eastern Ghats and Himalayas lie in the tropical/subtropical belt, their higher altitudes may provide a temperate environment at least during the winter season. Hence, the mycoflora of these tropical mountain streams often contains a small number of typically temperate species (Mer and Khulbe 1981; Madhusudan Rao and Manoharachary 1984; Manoharachary and Galiah 1987; Mer and Sati 1989; Sarma and Manoharachary 1989; Sati et al. 1989; Sridhar and Kaveriappa 1989a,d, 1992; Chandrashekar et al. 1990). A carefully designed study of aquatic hyphomycetes in locations where temperate and tropical climatic conditions overlap might provide clues on factors determining the occurrence of these rare species. It would complement Suberkropp's study (1984) of the occurrence of tropical species in a temperate climate.

Fungal colonization of substrata other than natural leaf litter is of considerable interest, especially where the natural forest has been replaced by large plantations. This has happened, for example, in large areas of the Western Ghats. Generally, aquatic hyphomycetes grow well on the litter of plantation crops, but, as is the case with tree leaves, there are differences between the various types of litter. The leaves of cocoa (*Theobroma cacao* Linn.), mango (*Mangifera indica* Linn.) and rubber decay more quickly than coffee, cashew and teak. An added complication is the potential impact of pesticides used to protect plantation crops. Inevitably, a certain proportion will reach the groundwater and streams, where they influence the fungi as well as leaf-eating invertebrates.

There have been few descriptions of new taxa occurring on the Indian subcontinent: *Triscelophorus konajensis* was isolated from submerged banyan leaves in a coastal stream (Sridhar and Kaveriappa 1987a), and *Flabellospora octacladia* (Saikia and Sarbhoy 1980) and *Speiropsis hyalospora* (Subramanian and Lodha 1964) were described from terrestrial sources. The scarcity of taxonomic work is partly due to the difficulty of isolating and maintaining cultures of aquatic hyphomycetes – they grow slowly, and heavy contamination by terrestrial fungi is common, especially during the monsoon and winter seasons (June–December). Mites and other terrestrial arthropods further contribute to the difficulty of maintaining pure cultures.

There has been a single published study on the effects of invertebrate feeding on the aquatic hyphomycete community in Indian waters. In dishes with incubated leaf litter, we have often observed partial consumption of conidia by unidentified aquatic invertebrates. How important are conidia as a food source for aquatic invertebrates? Are there animals that take advantage of the large numbers of conidia that accumulate in foam? How does the nutritional value of conidia differ from that of mycelium? In view of the huge numbers of conidia that are produced in streams (Chap. 3), these are potentially very important questions.

To date, aquatic hyphomycetes have been reported from only seven states in India, which account for approximately one third of the total area. Other ecological habitats which are of considerable interest include streams and rivers of semi-arid regions, of tropical rainforests, thermal springs in the Western Ghats and Himalayas, islands (Lakshdweep, Andamon and Nicobar), estuaries and the Thar desert. Reports on the occurrence of aquatic hyphomycetes are particularly valuable if they are combined with information on the substratum on which the fungus was growing, the season, water temperature and chemistry. This information would form a basis on which to proceed with physiological and biochemical studies.

Clavariopsis bulbosa is a common marine fungus. It was also reported in foam samples from streams in the Ananthagiri and Mannanur forests in the Eastern Ghats (Manoharachary and Murthy 1981). Possibly this fungus might serve as a good model to investigate factors that differentiate freshwater fungi from marine fungi.

Acknowledgements. We thank Prof. Felix Bärlocher for suggestions and helpful comments on the drafts of this chapter. Thanks are also due to B.S. Jayalakshmi for assistance in the survey of the literature and compilation of the bibliography. One of us (KRS) is thankful to Mangalore University for granting a study leave.

References

Alasoadura SO (1968a) Some aquatic hyphomycetes from Nigeria. Trans Br Mycol Soc 51:535–540

Alasoadura SO (1968b) *Flabellospora crassa* n. gen. n. sp. an aquatic hyphomycete from Nigeria. Nova Hedwigia 15:415–418

Alasoadura SO (1968c) *Flabellospora verticillata* a new species of aquatic hyphomycete from Nigeria. Nova Hedwigia 15:419–421

Anastasiou CJ (1964) Some aquatic fungi imperfecti from Hawaii. Pac Sci 18:202–222

Betancourt C, Caballero M (1983) Acuatic hyphomycetes (deuteromicotina) from Los Chorros, Utuado, Puerto Rico. Caribb J Sci 19:41–42

Betancourt C, Justiniano J (1989) Hongos ingoldianos presentes en el arroyo de Oro, Mayaguez, Puerto Rico. Int J Mycol Lichenol 4:1–8

Betancourt C, Cruz J, Garcia J, Galarza L (1986) Estudio preliminar de los hifomicetos acuaticos (deuteromicotina) de la Republica Dominicana. Caribb J Sci 22:49–51

Betancourt C, Cruz J, Garcia J (1987) Los hifomicetos acuaticos de la Quebrada Dona Juana en el Bosque Estatal de Toro Negro, Villalba, Puerto Rico. Caribb J Sci 23:278–284

Bhat DJ, Chien C-Y (1990) Water-borne hyphomycetes found in Ethiopia. Trans Mycol Soc Jpn 31:147–157

Chandrashekar KR (1987) Some aspects of water-borne fungi and their enzymes. Thesis, Mangalore University, Mangalore

Chandrashekar KR, Kaveriappa KM (1988) Production of extracellular enzymes by aquatic hyphomycetes. Folia Microbiol 33:55–58

Chandrashekar KR, Kaveriappa KM (1989) Effect of pesticides on the growth of aquatic hyphomycetes. Toxicol Lett 48:311–315

Chandrashekar KR, Kaveriappa KM (1991) Production of extracellular cellulase by *Lunulospora curvula* and *Flagellospora penicillioides*. Folia Microbiol 36:249–255

Chandrashekar KR, Sridhar KR, Kaveriappa KM (1986) Aquatic hyphomycetes of Kempu Hole in the Western Ghat forests of Karnataka. Indian Phytopathol 39:368–372

Chandrashekar KR, Sridhar KR, Kaveriappa KM (1989) Palatability of rubber leaves colonized by aquatic hyphomycetes. Arch Hydrobiol 115:361–369

Chandrashekar KR, Sridhar KR, Kaveriappa KM (1990) Periodicity of water-borne hyphomycetes in two streams of Western Ghat forests (India). Acta Hydrochim Hydrobiol 18:187–204

Chandrashekar KR, Sridhar KR, Kaveriappa KM (1991) Aquatic hyphomycetes of a sulphur spring. Hydrobiologia 218:151–156

Cowling SW, Waid JS (1963) Aquatic hyphomycetes in Australia. Aust J Sci 26:122–123

Crane JL, Dumont KP (1975) Hyphomycetes from the West Indies and Venezuela. Can J Bot 53:843–851

Dixon PA (1959) Stream spora in Ghana. Trans Br Mycol Soc 42:174–176

Galiah K, Manoharachary C (1987) Studies on conidial fungi of a stream from Andhra Pradesh. Indian Phytopathol 40:446–473

Goos RD (1970) In vitro sporulation in *Actinospora megalospora*. Trans Br Mycol Soc 55:335–337

Goos RD (1978) Occurrence of *Triscelophorus monosporus* in upland sites on Oahu, Hawaii. Mycologia 70:188–189

Hasija SK, Shanware V (1986) Aquatic hyphomycetes of India. I. Two new records. Proc Natl Acad Sci India 56 B:74–76

Hudson JH (1961) *Heliscus submersus* sp. nov. an aquatic hyphomycete from Jamaica. Trans Br Mycol Soc 44:91–94

Hudson JH, Ingold CT (1960) Aquatic hyphomycetes from Jamaica. Trans Br Mycol Soc 43:469–478

Ingold CT (1942) Aquatic hyphomycetes of decaying alder leaves. Trans Br Mycol Soc 25:339–417

Ingold CT (1956) Stream spora in Nigeria. Trans Br Mycol Soc 39:106–110

Ingold CT (1958) Aquatic hyphomycetes from Uganda and Rhodesia. Trans Br Mycol Soc 41:109–114

Ingold CT (1959) Aquatic spora of Omo forest, Nigeria. Trans Br Mycol Soc 42:479–485

Ingold CT (1960) Aquatic hyphomycetes in southern Rhodesia. Proc Rhodesian Sci Assoc 43:49–53

Ingold CT (1973) Branched aquatic propagules with four or five divergent arms. Acta Bot Indica 1:69–72

Ingold CT (1975) An Illustrated Guide to Aquatic and Water-borne Hyphomycetes (Fungi Imperfecti) with notes on their Biology. Freshwater Biol Assoc Sci Pub No 30, pp 96

Ingold CT (1976) The morphology and biology of freshwater fungi excluding phycomycetes. In: Jones EBG (ed) Recent advances aquatic mycology. Elek Science, London, pp 335–357

Ingold CT, Webster J (1973) Some aquatic hyphomycetes from India. Kavaka 1:5–9

Iqbal SH, Webster J (1973) Aquatic hyphomycete spora of River Exe and its tributaries. Trans Br Mycol Soc 61:331–336

Kuthubutheen AJ (1987) A new species of *Phalangispora* and further observations on *P. constricta* from Malaysia. Trans Br Mycol Soc 89:414–420

Kuthubutheen AJ, Nawawi A (1987) A new species of *Speiropsis* from Malaysia. Trans Br Mycol Soc 89:584–587

Kuthubutheen AJ, Nawawi A (1988) A new species of *Wiesneriomyces* (hyphomycetes) from submerged decaying leaves. Trans Br Mycol Soc 90:619–625

Le'John HB (1965) Sierra Leone freshwater hyphomycetes. Trans Br Mycol Soc 48:261–264

Madhusudan Rao M, Manoharachary C (1982) Processing of leaves by fungi in an aquatic ecosystem. Indian Phytopathol 35:654–656

Madhusudan Rao M, Manoharachary C (1984) Taxo-ecological studies on some aquatic hyphomycetes from India. Indian Phytopathol 37:64–68

Manoharachary C (1989) Some aspects of water-borne conidial fungi from India. In: Mukerji KG, Singh VP, Garg KL (eds) Forntiers in Applied Microbiology vol 3, Rastogi, Meerut, India, pp 197–208

Manoharachary C, Bhairavanath D (1985) Conidial fungi from freshwater foam. Indian J Bot 8:67–68

Manoharachary C, Galiah K (1987) Studies on conidial fungi associated with foam and submerged leaves of a stream in Andhra Pradesh. In: Hasija SK, Rajak RC, Singh SM (eds) Perspectives in mycological research, vol 1, Today and Tomorrow's Printers and Publishers, New Delhi, pp 69–77

Manoharachary C, Madhusudan Rao M (1983) Ecological studies of hyphomycetes associates with submerged leaves from India. Indian Phytopathol 36:62–65

Manoharachary C, Murthy ABS (1981) Foam spora from Andhra Pradesh, India. Curr Sci 50:378–379

Marvanová L, Marvan P (1969) Aquatic hyphomycetes in Cuba. Ceska Mycol 23:135–140

Matsushima T (1971a) Mycological reports from New Guinea and the Solomon Islands. 7. Some interesting Fungi imperfecti. Bull Natl Sci Mus Tokyo 14:460–480

Matsushima T (1971b) Microfungi of the Solomon Islands and Papua New Guinea. Shionogi Res Lab, Shionogi and Co Ltd, Kobe, pp 78

Matsushima T (1980) Saprophytic microfungi from Taiwan, part I. Hyphomycetes. Matsushima Mycological Memoirs No 2:1–82

Mer GS, Khulbe RD (1981) Aquatic hyphomycetes of Kumaun Himalaya, India. Sydowia 34:118–124

Mer GS, Sati SC (1989) Seasonal fluctuations in species composition of aquatic hyphomycetetous flora in a temperate freshwater stream of Central Himalaya, India. Int Revue Gesamten Hydrobiol 74:433–437

Nawawi A (1985a) Aquatic hyphomycetes and other water-borne fungi from Malaysia. Malays Nat J 39:75–134

Nawawi A (1985b) Some interesting hyphomycetes from water. Mycotaxon 24:217–226

Nawawi A (1987) *Clavariopsis azlanii* sp. nov. a new aquatic hyphomycete from Malaysia. Trans Br Mycol Soc 88:428–432

Nawawi A, Kuthubutheen AJ (1987) *Tetrabrachium elegans* gen. and sp. nov. a hyphomycete with tetraradiate conidia. Mycotaxon 24:291–297

Nawawi A, Kuthubutheen AJ (1988a) *Tricladiospora* a new genus of dematiaceous hyphomycetes with staurosporus conidia from submerged decaying leaves. Trans Br Mycol Soc 90:482–487

Nawawi A, Kuthubutheen AJ (1988b) *Tricladiomyces geniculatus* sp. nov. a conidial basidiomycete. Trans Br Mycol Soc 90:670–673

Nawawi A, Kuthubutheen AJ (1990) *Nidulispora* gen. nov. a hyphomycete genus with crateriform conidia. Mycotaxon 36:329–336

Nayo SG (1975) The ecology and distribution of aquatic hyphomycetes around Bogor. Biotrop, SEAMEO Regional Centre for Tropical Biology, TFRS 75/151, Bogor, Indonesia, pp 61

Nilsson S (1962) Some aquatic hyphomycetes from South America. Sven Bot Tidskr 56:351–361

Nilsson S (1964) Freshwater hyphomycetes. Symb Bot Ups 18:1–130

Padgett D (1976) Leaf decomposition by fungi in a tropical rainforest stream. Biotropica 8:166–178

Patil SD, Kapadnis BP (1980) Stream spora of Maharashtra, India. MVM Patrika 14:59–64

Patil AS, Rao VG (1972) *Articulospora* – a new record from India. Indian Phytopathol 25:455–456

Ranzoni FV (1979) Aquatic hyphomycetes from Hawaii. Mycologia 71:786–795

Saikia UN, Sarbhoy AK (1980) Hyphomycetes of North East India I. Indian Phytopathol 33:459–461

Sarma TK, Manoharachary C (1989) Numerical studies on conidial fungi from a stream of Andhra Pradesh. Indian Phtopathol 42:596–598

Sati SC, Tiwari N (1990) Some aquatic hyphomycetes of Kumaun Himalaya, India. Mycotaxon 39:407–414

Sati SC, Mer GS, Tiwari N (1989) Occurrence of water-borne conidial fungi on *Pinus Roxburghii* needles. Curr Sci 58:918

Shaw D (1972) *Ingoldiella hamata* gen. et sp. nov. a fungus with clamp connexions from a stream in North Queensland. Trans Br Mycol Soc 59:255–259

Shaw D, Sutton BC (1985) A new aero-aquatic hyphomycete from Papua New Guinea and Australia. Bot J Linn Soc 91:25–36
Shearer CA, Lane LC (1983) Comparision of three techniques for the study of aquatic hyphomycete communities. Mycologia 75:498–508
Singh N (1972) *Pyramidospora constricta* sp. nov. a new aquatic hyphomycete. Trans Br Mycol Soc 59:336–339
Singh N (1976) *Pyramidospora herculiformis* sp. nov. a new aquatic hyphomycete from Sierra Leone. Trans Br Mycol Soc 66:347–350
Singh N, Musa TM (1977) Terrestrial occurrence and the effect of temperature on growth, sporulation and spore germination of some tropical aquatic hyphomycetes. Trans Br Mycol Soc 68:103–105
Sridhar KR (1984) Studies on water-borne fungi of Dakshina Kannada and Kodagu regions. Thesis, Mangalore University, Mangalore
Sridhar KR, Kaveriapppa KM (1982) Aquatic fungi of the Western Ghat forests in Karnataka. Indian Phytopathol 35:293–296
Sridhar KR, Kaveriappa (1983) Culturing water-borne hyphomycetes on plant petioles. Indian Bot Rep 2:111–115
Sridhar KR, Kaveriappa KM (1984a) Aquatic hyphomycetes of Western Ghat forests in Karnataka. Indian Phytopathol 37:546–548
Sridhar KR, Kaveriappa KM (1984b) Seasonal occurrence of water-borne fungi in Konaje stream (Mangalore), India. Hydrobiologia 119:101–105
Sridhar KR, Kaveriappa KM (1985a) Water-borne fungi of Kunthi river in Silent Valley, Kerala. Indian Phytopathol 38:371–372
Sridhar KR, Kaveriappa KM (1985b) Water-borne hyphomycetes of the Western Ghat forests. Indian Phytopathol 38:557–558
Sridhar KR, Kaveriappa KM (1986a) New host records of aquatic hyphomycetes. Indian Phytopathol 39:131–132
Sridhar KR, Kaveriappa KM (1986b) Effect of pesticides on sporulation and spore germination of water-borne hyphomycetes. In: Dalella RC, Madhyastha MN, Joseph MM (eds) Environmental biology-coastal ecosystem. The Academy of Environmental Biology, India, pp 195–204
Sridhar KR, Kaveriappa KM (1987a) A new species of *Triscelophorus*, Indian Phytopathol 40:102–105
Sridhar KR, Kaveriappa KM (1987b) Culturing water-borne hyphomycetes on plant latex. J Indian Bot Soc 66:232–233
Sridhar KR, Kaveriappa KM (1987c) Occurrence and survival of aquatic hyphomycetes under terrestrial conditions. Trans Br Mycol Soc 89:606–609
Sridhar KR, Kaveriappa KM (1988a) New host records of aquatic hyphomycetes. Indian Phytopathol 41:160–161
Sridhar KR, Kaveriappa KM (1988b) Colonization of leaf litter by aquatic hyphomycetes in Western Ghat stream. Proc Indian Natl Sci Acad B 54:199–200
Sridhar KR, Kaveriappa KM (1988c) Colonization of leaf litter by aquatic hyphomycetes in a tropical stream. Arch Hydrobiol 112:627–630
Sridhar KR, Kaveriappa KM (1988d) Occurrence and survival of aquatic hyphomycetes in brackish and seawater. Arch Hydrobiol 113:153–160
Sridhar KR, Kaveriappa KM (1988e) Survival of water-borne fungi-imperfecti under non-aquatic conditions. Proc Indian Natl Sci Acad B 54:295–297
Sridhar KR, Kaveriappa KM (1989a) Notes on aquatic hyphomycetes of mountain streams in Western Ghat region, India. Feddes Repert 100:187–189
Sridhar KR, Kaveriappa KM (1989b) Colonization of leaves by water-borne hyphomycetes in a tropical stream. Mycol Res 92:392–396
Sridhar KR, Kaveriappa KM (1989c) Water-borne hyphomycetes spora of two freshwater streams. Environ Ecol 7:771–772
Sridhar KR, Kaveriappa KM (1989d) Observations on aquatic hyphomycetes of the Western Ghat streams, India. Nova Hedwigia 42:455–467
Sridhar KR, Kaveriappa KM (1989e) New substrates of aquatic hyphomycetes. Indian Phytopathol 42:203

Sridhar KR, Kaveriappa KM (1990) Water-borne hyphomycetes of two streams in Dakshina Kannada. Indian Phytopathol 43 (in press)
Sridhar KR, Kaveriappa KM (1992) Aquatic hyphomycetes of Western Ghat streams, India. Sydowia 44 (in press)
Suberkropp K (1984) Effect of temperature on seasonal occurrence of aquatic hyphomycetes. Trans Br Mycol Soc 82:53–62
Subramanian CV, Bhat DJ (1981) Conidia from freshwater foam samples from the Western Ghats, Southern India. Kavaka 9:45–62
Subramanian CV, Lodha BC (1964) Two interesting hyphomycetes. Can J Bot 42:1057–1063
Swart HJ (1986) Preliminary survey of aquatic hyphomycetes in Victoria, Australia. Trans Br Mycol Soc 86:497–501
Thakur SB (1977) Survival of some aquatic hyphomycetes under dry conditions. Mycologia 69:843–845
Thomas K, Chilvers GA, Norris RH (1989) Seasonal occurrence of conidia of aquatic hyphomycetes (fungi) in Lees Creek, Australian Capital Territory. Aust J Mar Freshwater Res 40:11–23
Tubaki K (1965) Short note on aquatic spora in East New Guinea. Trans Mycol Soc Jpn 6:11–16
Tubaki K, Watanabe K, Manoch L (1983) Aquatic hyphomycetes from Thailand. Trans Mycol Soc Jpn 24:451–457
Webster J, Descals E (1981) Morphology, distribution and ecology of conidial fungi in freshwater habitats. In: Cole GT, Kendrick B (eds) The biology of conidial fungi. Academic Press, New York, pp 295–355

11 Conclusions and Outlook

Felix Bärlocher

The shedding of plant parts has a long history. Branch abscission was found among Lycopsida in the Paleozoic (Millingot and Chaney 1973). The earliest fossil record of deciduous leaves in members of the Glosslopteridae in the Southern Hemisphere dates to the Carboniferous, some 300 million years ago. They shed leaves in clusters on short shoots, much like needle clusters of present day *Pinus* are shed (Addicott and Lyon 1973). The deciduous habit of angiosperm trees of the Northern Hemisphere developed during the early Cretaceous, over 100 million years ago. In both hemispheres, the deciduous habit evolved in conjunction with the establishment of a strongly seasonal climate, indicating that ecological factors had a decisive influence on its evolution.

Most major freshwater invertebrate groups originated from marine ancestors. Only a few groups, such as insects (but see below), mites, and pulmonate snails are presumed to have clearly originated from terrestrial habits (Pennak 1978). Today, the major macroinvertebrate leaf shredders belong to the insects, and to a lesser degree, to amphipods and isopods. The fossil record of amphipods is meagre, extending back only to the Eocene; isopods are known from the Triassic (or Permian) (Bousfield 1973).

Wootton (1988) discussed the recent view that postulate an aquatic or amphibiotic stage in the ancestry of all insects. If correct, this would have profound implications: insects would have been a component of freshwater systems since the Silurian or the Devonian. Wootton (1988) concluded that this view is neither confirmed nor disproved by the observed distribution of freshwater insects in time. The fossil record does suggest that running waters were colonized before lentic waters. Many early insects were predatory. Trichoptera larvae with cases were probably the first significant insect shredders of coarse plant debris. They originated in the Jurassic-Cretaceous border, coincident with the rise of the angiosperms. Thus, the association between leaves and stream shredders is over 100 million years old.

Buchner (1965), who did much of the early work on invertebrate/microbe symbiosis, noticed that endosymbiotic bacteria and yeasts are much less common in aquatic than in terrestrial invertebrates. He postulated that the aquatic environment generally provides better growth conditions for microorganisms; consequently, the animals should have no problem satisfying their need for microbial vitamins, proteins, or enzymes by selectively ingesting adequately colonized substrata. Though this idea has never been seriously tested, it seems plausible. In streams, the shredders with the richest intestinal microflora are larvae of *Tipula* (Chap. 6). Almost without exception, a leaf collected from a stream carries some aquatic hyphomycetes. It also provides a substratum for epiphyllic bacteria. An important qualification must be made, however: the nutritional quality of decomposing leaves does not improve indefinitely due to microbial conditioning. Within a few weeks, a peak is reached; continued microbial activity will then lead to a deterioration of the food (Chap. 6).

How long ago did this microbial link between plant litter and invertebrates evolve? Bacteria, of course, were around much longer than either of the other two groups, but what about the fungi? Terrestrial anamorphs we know today have changed little since the Tertiary period and possibly since much earlier times (Pirozynski and Weresub 1979). There are few unambiguous records of freshwater fungi (Sherwood-Pike 1988), and no records of unequivocally freshwater sexual stages of Ascomycetes or Basidiomycetes. Some helicoid spores were found from the Cretaceous and Tertiary (but not necessarily from aquatic fungi). Conidia of two extant genera of aquatic hyphomycetes, plus other spores whose morphology suggested an aquatic origin, were found in Miocene lake bed assemblages. Many of these spores may have been washed in from surrounding streams. Most aquatic hyphomycetes have thinwalled and colourless conidia, which are therefore unlikely to have been preserved. Sherwood-Pike (1988) suggests that a survey of extant fungi in lakes using palynological methods might provide useful comparative data. In addition, recently developed techniques in molecular biology will undoubtedly throw some light on the question of how long ago aquatic hyphomycetes separated from their terrestrial ancestors.

Of all aspects of the ecology of aquatic hyphomycetes, their role as intermediaries between leaves and invertebrates has attracted most attention outside mycology. This is hardly surprising, since, through their effect on invertebrate feeding, the fungi may ultimately influence higher trophic levels such as fish. It is encouraging that fungi rated a separate food category in a recent analysis of invertebrate gut contents of a blackwater stream (Wallace et al. 1987). Twenty years ago, most stream ecologists would have been unaware of the potential significance of the fungi. Wallace et al. (1987) estimated their contribution to invertebrate diets to be a few percent. Though there are several problems associated with gut analyses (e.g. the difficulty of reliably measuring mycelial fragments; the connection between recognizable food items in the gut and their contribution to the diet, etc.), this was an important first step toward quantifying the actual fungal significance in the field. Laboratory experiments tend to provide the best possible conditions for fungal conditioning (though some investigators are less successful in this than others; see Chap. 6). To justify this, it is often assumed that selective feeding can make up for uneven distribution or quality of food items in streams. I think it would be premature to accept this as a fact. Leaves collected from a stream varied considerably in their protein content and digestibility (Bärlocher 1983). Trade-offs between ingesting optimally conditioned food, search time and avoidance of predators may have to be considered. There is little evidence at this point that sweeping generalizations are useful. A case by case approach may be necessary.

This brings me to a point emphasized, among others, by Suberkropp (Chap. 6) and Chamier (Chap. 8): research on aquatic hyphomycetes has reached a stage where conclusions become increasingly qualified, and more and more effort has to be spent on defining and standardizing experimental conditions. A researcher would no longer claim to have established the effects of fungi on stream invertebrates, but, for example, to have determined the growth rate of three week old *Gammarus pseudolimnaeus* kept at 10 °C, feeding on a diet of *Acer saccharum*, inoculated with a strain of *Tricladium angulatum*, and incubated for two weeks at 10 °C in a defined mineral solution. Similarly, to determine connections between aquatic hyphomycetes and water chemistry, more and more detailed analyses seem desirable, and conclusions become more and more restricted and conditional. This development is unavoidable; broad generalization from the early, exciting period are now seen as special cases in a vast array of possibilities. Final answers seem to recede in a blur of ever finer detail. These refinements are not without danger: hypotheses may be formulated too narrowly, and we might miss the unexpected pattern.

Two very important recent developments are the examination of unusual substrata and sites for the occurrence of aquatic hyphomycetes, and connected with it, sexual reproduction (Chaps. 2, 4, 5). The significance of sexual reproduction in the ecology of aquatic hyphomycetes remains unknown. A combination of field observations and modern techniques of molecular biology (Trevors and van Elsas 1989) may provide information on this topic, which remains of central interest to evolutionary theory (Bell 1982; Stearns 1987). The same techniques may also prove helpful to determine the validity of current taxonomic concepts. Classification of the anamorphs is based on spore morphology and development.

Until quite recently, it was widely assumed that conspecific mycelia that happen to encounter each other in the field would fuse and form a unit (Burnett 1976). At least since the publications by Williams (1966) and Dawkins (1976), the inherently "selfish" behaviour of individuals or genes makes such far-reaching cooperation unlikely. Rayner and Todd (1977) therefore proposed the concept of "individualistic mycelia". These maintain their genetic integrity by somatic incompatability, which generally expresses itself by a demarcation zone of distinctive mycelium (e.g. sparse mycelium, pigmented zone, aerial mycelium). In the only study of aquatic hyphomycetes dealing with this topic, no signs of somatic incompatability between strains isolated from widely separated locations were detected (Bärlocher 1991). This does not necessarily indicate frequent genetic exchange. It is difficult to imagine which benefits the nuclei of one mycelium might gain from admitting nuclei of another mycelium and thereby sharing the protoplasm used for conidium formation. Nevertheless, genetic exchanges can occur when non-identical mycelia meet (Rayner 1991). The strength of the incompatability reaction appears to vary with the functional mode of the mycelium, suggesting that the costs/benefits of mycelial fusion may vary (Stenlid and Rayner 1989). The remarkable similarity of aquatic hyphomycete species from different continents suggest that genetic exchange does occur. Techniques of molecular biology may elucidate the processes involved.

One of the real surprises in recent years has been the discovery and documentation of extensive mucilage production on conidia and hyphae of aquatic hyphomycetes (Chap. 7). It obviously plays a crucial role in the initial stages of colonization; possibly, it is also involved in enzyme delivery or immobilization. This topic will undoubtedly continue to yield fascinating results.

Finally, a correction of the geographic bias, i.e. the restriction to temperate regions, is long overdue. I'm convinced that many of our favourite assumptions concerning the biology of aquatic hyphomycetes will have to be readjusted, as more reports from tropical regions are becoming available. Sridhar et al. report the intriguing observation that conidia were being eaten by invertebrates (Chap. 10). In view of the enormous numbers produced, this could be a very important food source for stream organisms.

There is little doubt that impressive advances have been made in our knowledge of the taxonomy, physiology and ecology of aquatic hyphomycetes since they were discovered by Ingold in 1942. But to workers active in a field, it is inevitably the unknown areas that tend to dominate their attention.

Blaise Pascal once compared knowledge to a sphere. As our knowledge (the sphere) grows, so does our contact with the unknown (the sphere's surface). But, due to the relationship between surfaces and volumes, the larger the sphere, the greater the ratio of known (volume) to unknown (surface). Gould (1980) concludes one of his essays with the following wish, which also seems appropriate for the ecology of aquatic hyphomycetes: "May absolutely increased ignorance continue to flourish with relatively increased knowledge."

References

Addicott FT, Lyon JL (1973) Physiological ecology of abscission. In: Kozlowski TT (ed) Shedding of plant parts, Academic Press, New York, pp 85–124

Bärlocher F (1983) Seasonal availability and digestibility of CPOM in a stream. Ecology 64:1266–1272

Bärlocher F (1991) Intraspecific hyphal interactions among aquatic hyphomycetes. Mycologia 83:82–88
Bell G (1982) The masterpiece of nature – the evolution and genetics of sexuality. University of California Press, Berkeley, California
Bousfield EL (1973) Shallow-water gammaridean amphipoda of New England. Cornell University Press, Ithaca and London
Buchner K (1965) Endosymbiosis of animals with plant microorganisms. Wiley, New York
Burnett JH (1976) Fundamentals of mycology. Arnold, London
Dawkins R (1976) The selfish gene. Oxford University Press, Oxford
Gould SJ (1980) Ever since Darwin. Penguin, New York
Ingold CT (1942) Aquatic hyphomycetes of decaying alder leaves. Trans Br Mycol Soc 25:339–417
Millingot WF, Chaney WR (1973) Shedding of shoots and branches. In: Kozlowski TT (ed) Shedding of plant parts. Academic Press, New York, pp 149–204
Pennak RW (1978) Freshwater invertebrates of the United States. Wiley-Interscience, New York
Pirozynski KA, Weresub LK (1979) The classification and nomenclature of fossil fungi. In: Kendrick B (ed) The whole fungus. National Museums of Canada, Ottawa, vol 2, pp 653–688
Rayner ADM (1991) The challenge of the individualistic mycelium. Mycologia 83:48–71
Rayner ADM, Todd NK (1977) Intraspecific antagonism in natural populations of wood-decaying basidiomycetes. J Gen Microbiol 103:85–90
Sherwood-Pike MA (1988) Freshwater fungi: fossil record and paleoecological potential. Palaeogeogr Palaeoclimatol Palaeoecol 62:271–285
Stearns SC (ed) (1987) The evolution of sex and its consequences. Birkhäuser, Basel
Stenlid J, Rayner ADM (1989) Environmental and endogenous controls of developmental pathways: variation and its significance in the forest pathogen, *Heterobasidium annosum*. New Phytol 113:245–258
Trevors JT, van Elsas JC (1989) A review of selected methods in environmental microbial genetics. Can J Microbiol 35:895–902
Wallace JB, Benke AC, Lingle AH, Parsons K (1987) Trophic pathways of macroinvertebrate primary consumers in subtropical blackwater streams. Arch Hydrobiol (Suppl) 74:423–451
Williams GC (1966) Adaptation and natural selection. Princeton University Press, Princeton
Wootton RJ (1988) The historical ecology of aquatic insects: an overview. Palaeogeogr Palaeoclimatol Palaeoecol 62:477–492

Subject Index

Printing: Druckerei Kutschbach, Berlin
Binding: Buchbinderei Lüderitz & Bauer, Berlin